INDUSTRY'S GUIDE TO ISO 9000

ETM WILEY SERIES IN ENGINEERING & TECHNOLOGY MANAGEMENT

Series Editor: Dundar F. Kocaoglu, Portland State University

Badiru/ INDUSTRY'S GUIDE TO ISO 9000

Badiru/ PROJECT MANAGEMENT IN MANUFACTURING AND HIGH TECHNOLOGY OPERATIONS

Baird/ MANAGERIAL DECISIONS UNDER UNCERTAINTY: AN INTRODUCTION TO THE ANALYSIS OF DECISION MAKING

Edosomwan/ INTEGRATING INNOVATION AND TECHNOLOGY MANAGEMENT

Eschenbach/ CASES IN ENGINEERING ECONOMY

Gerwin and Kolodny/ MANAGEMENT OF ADVANCED MANUFACTURING TECHNOLOGY: STRATEGY, ORGANIZATION, AND INNOVATION

Gönen/ ENGINEERING ECONOMY FOR ENGINEERING MANAGERS

Jain and Triandis/ MANAGEMENT OF RESEARCH AND DEVELOPMENT ORGANIZATIONS: MANAGING THE UNMANAGEABLE

Lang and Merino/ THE SELECTION PROCESS FOR CAPITAL PROJECTS

Martin/ MANAGING INNOVATION AND ENTREPRENEURSHIP IN TECHNOLOGY-BASED FIRMS

Martino/ R & D PROJECT SELECTION

Messina/ STATISTICAL QUALITY CONTROL FOR MANUFACTURING MANAGERS

Morton and Pentico/ HEURISTIC SCHEDULING SYSTEMS: WITH APPLICATIONS TO PRODUCTION SYSTEMS AND PROJECT MANAGEMENT

Niwa/ KNOWLEDGE BASED RISK MANAGEMENT IN ENGINEERING: A CASE STUDY IN HUMAN-COMPUTER COOPERATIVE SYSTEMS

Porter et al./ FORECASTING AND MANAGEMENT OF TECHNOLOGY

Riggs/ FINANCIAL AND COST ANALYSIS FOR ENGINEERING AND TECHNOLOGY MANAGEMENT

Rubenstein/ MANAGING TECHNOLOGY IN THE DECENTRALIZED FIRM

Sankar/ MANAGEMENT OF INNOVATION AND CHANGE

Streeter/ PROFESSIONAL LIABILITY OF ARCHITECTS AND ENGINEERS

Thamhain/ ENGINEERING MANAGEMENT: MANAGING EFFECTIVELY IN TECHNOLOGY-BASED ORGANIZATIONS

INDUSTRY'S GUIDE TO ISO 9000

ADEDEJI BODUNDE BADIRU

A WILEY-INTERSCIENCE PUBLICATION
JOHN WILEY & SONS, INC.
NEW YORK CHICHESTER BRISBANE TORONTO SINGAPORE

Coventry University

This text is printed on acid-free paper.

Copyright © 1995 by John Wiley & Sons, Inc.

All rights reserved. Published simultaneously in Canada.

Reproduction or translation of any part of this work beyond
that permitted by Section 107 or 108 of the 1976 United
States Copyright Act without permission of the copyright
owner is unlawful. Requests for permission or further
information should be addressed to the Permissions Department,
John Wiley & Sons, Inc., 605 Third Avenue, New York, NY
10158-0012.

This publication is designed to provide accurate and
authoritative information in regard to the subject
matter covered. It is sold with the understanding that
the publisher is not engaged in rendering legal, accounting,
or other professional services. If legal advice or other
expert assistance is required, the services of a competent
professional person should be sought.

Library of Congress Cataloging-in-Publication Data:
Badiru, Adedeji Bodunde, 1952–
    Industry's guide to ISO 9000 / Adedeji Bodunde Badiru.
      p. cm.
    Includes index.
    ISBN 0-471-04598-5
    1. ISO 9000 Series Standards.   I. Title.
TS156.6.B34   1995
658.5'62—dc20   94-48504

Printed in the United States of America

10 9 8 7 6 5 4 3 2 1

To my wife, Iswat, whose love sustains me in everything that I do.

CONTENTS

Preface   xiii

Acknowledgments   xv

1. Global Quality Perspectives   1
   Global Competition, 2
   Definitions of Quality, 2
   The Need for Standards, 3
   International Standards, 3
   Impact of Trade Agreements, 4
   Quality Awards, 5
   Preparing for the Change, 6
   Planning Levels, 6
   Teamwork, 7
   Benchmarking for ISO 9000, 8
   Management Leadership, 9
      Upper Management Responsibilities, 10
   Risk Management, 13
   ISO 9000 Mission Logistics, 14

2. Introduction to ISO 9000   17
   Quality Proclamation, 17
   What is ISO 9000? 17

What's in the Name? 19
   What about Terminology? 20
Subdivisions of ISO 9000, 20
Purpose of ISO 9000, 21
What Is the Quality System? 22
Acceptance of ISO 9000, 25
   Benefits of ISO 9000 Registration, 26
Getting Started, 27
   Major Phases of ISO 9000 Registration Process, 28
Internal Audit, 29
   Checklist, 29
Third-Party Audit, 31
   Third-Party Assessment Steps, 32
Documentation Steps, 33
   Documentation Process Summary, 34
   Documentation Survey, 36
   Paper Audit, 38
   Employee Job Descriptions, 40
Site Implementation Plan, 41
ISO Policy and Procedures, 41
A Twelve-Month Timeline, 42
Tips for a Successful Process, 42
Impact of Technology Changes, 43

3. TQM and ISO 9000 — 45

TQM System, 45
   Benefits of TQM, 47
Just in Time for ISO 9000, 47
Components of Quality Improvement, 48
Customer Involvement, 48
   Producer-Consumer Partnership, 49
   User Training and Instructions, 51
Vendor Certification, 51
   Vendor Rating System, 52
Employee Involvement, 54
Quality of Manufactured Goods, 55
   Managing Product Complexity, 55
Quality of Service, 57
Management by Objective, 58
   Management by Exception, 58

Ongoing Commitment, 59
Prevention versus Cure, 60
Continuous Process Improvement, 60
   Pitfall of Ambiguous Customer Surveys, 64
Case Study: TQM in Soccer? 65

4. ISO 9001 Quality System Requirements   67

Management Responsibility, 70
   Quality Policy, 70
   Organization, 70
Quality System Documentation, 71
Contract Review, 72
Design Control, 72
Document Control, 73
Purchasing, 74
Purchaser-Supplied Product, 74
Product Identification and Traceability, 75
Process Control, 75
Inspection and Testing, 76
Inspection, Measuring, and Test Equipment, 77
Inspection and Test Status, 78
Control of Nonconforming Product, 78
Corrective Action, 79
Handling, Storage, Packaging, and Delivery, 79
Quality Records, 80
Internal Quality Audits, 80
Training, 80
Servicing, 81
Statistical Techniques, 81

5. ISO 9000 Project Management   83

Elements of Project Management, 83
Steps of Project Management, 86
   Overview, 86
Selecting the Project Manager, 90
Selling the Project Plan, 91
Staffing the Project, 92
ISO 9000 Project Decision Process, 94
   Project Decision Steps, 94
Conducting ISO 9000 Meetings, 97

Group Decision Making, 100
   Various Methods, 101
Personnel Management, 106
Systems Integration for ISO 9000, 107
Project Blueprint, 109
   Outline, 110

6. ISO 9000 Planning Guidelines   115
Phases of Preparation, 115
Strategic Planning, 116
Time-Cost-Performance Criteria, 117
Components of an ISO 9000 Plan, 120
Employee Motivation, 122
   Various Approaches, 122
Feasibility Study, 126
Budget Planning, 128
   Top-Down Budgeting, 128
   Bottom-Up Budgeting, 129
Work Breakdown Structure, 129
Information Flow for ISO 9000, 131
Triple C Approach to Planning, 134
   Communication, 135
   Cooperation, 140
   Coordination, 142
   Conflict Resolution, 145

7. Quality Audit Techniques   149
Purpose of Internal Audit, 149
Types of Audits, 150
   System Audit, 150
   Process Audit, 151
   Product Audit, 151
Management Review, 151
Audit Team Preparation, 152
Sampling Strategy, 152
   Data Measurement Scales, 153
   Sample Characteristics, 154
Interviewing, 157
The Paper Audit, 159
   Questions, 159

Is This the End of the Journey? 168

Appendix A: Sources of ISO 9000 Information and Resources 169

Appendix B: Glossary of International Business Terms 173

Appendix C: Glossary of Quality-Related Terms 185

Appendix D: Units and Measures Conversion Factors 199

Selected Bibliography 205

Index 209

PREFACE

Quality is a universal language. The existence of standards helps to achieve consistent levels of product quality. The emergence of ISO 9000 has created the need to refocus the way companies operate. Customers and consumers are so sophisticated now that they will no longer simply accept whatever is offered in the market. In the past, consumers were expected to make do with the inherent quality of the available product, no matter how poor it was. This has changed drastically in the past few years. For a product to satisfy the sophisticated taste of the modern consumer, it must exhibit a high level of quality. Only high-quality products and services can survive the prevailing market competition.

This book presents practical guidelines for pursuing the requirements of ISO 9000. This book is not about the standards themselves, but rather about the process needed to comply with the standards. The primary audience for the book consists of practitioners in all functional areas of business and industry. Examples of the relevant functional areas include industrial and systems engineers, process engineers, designers, R & D managers, plant managers, production supervisors, manufacturing engineers, and quality engineers. The book should also appeal to academic institutions and professional training organizations as a reference material.

The book also considers the human aspects of pursuing ISO 9000. Regardless of the technical capabilities of the production machinery of a company, it is people that will make the quality goals of a company realizable. The book is written in a concise and clear language suitable for easy reference. Guidelines are presented for the various aspects of ISO 9000 and the organizational processes required. The project management guidelines presented in Chapters Five and Six are essential for successfully carrying out the planning, organizing, scheduling, and control functions that support the ISO 9000 process.

ISO 9000 deals mostly with people-related issues, rather than technical systems. The usual process, if not managed well, can be so arduous that people are just glad when it is over. But the truth is, it should never be over. It should be an ongoing process. Maintaining what has been achieved is as important as achieving it in the first place.

Thus, the unique aspect of this book is the use of project management techniques to manage the ISO 9000 process. With proper management, even an arduous task can become pleasant. It is recommended that the entire book be read so that specific topics and tools of interest can be extracted and organized into a custom certification plan to satisfy the needs of the organization.

ADEDEJI BODUNDE BADIRU

ACKNOWLEDGMENTS

I greatly thank Matt P. Jung and John Best of Seagate Technology for sharing their ISO 9000 experience for the preparation of this manuscript. I also thank Terry Judah and Tim Mendenhall of Hitachi Computer Products (America), Inc., for their helpful comments and suggestions after reviewing the initial draft of this manuscript. My thanks also go to Dave Chandler and Dr. Ghassan Abdelnour of Seagate Technology for their continuing friendship, support, and encouragement, and for providing practical insights into the real world of product-quality issues. I should also express my gratitude to Dr. Mustapha Pulat of AT&T, whose comments in a casual conversation initiated the idea for this book. My thanks to the editorial and production staff at John Wiley. Thanks to Frank J. Cerra, former editor at John Wiley, for his visionary interest in this book right from the beginning and to Bob Argentieri for picking up editorial interest in the book expeditiously. As always, I thank my colleagues in the School of Industrial Engineering, University of Oklahoma, for their continuing support. Special thanks go to the secretarial staff in the School of Industrial Engineering, Jane Smith, Lisa Robinett, and Jean Shingledecker, for their commitment to high-quality service and dedication to duty. Finally, I thank my wife, Iswat, for continuing to create a conducive atmosphere in which I can think and write.

1
GLOBAL QUALITY PERSPECTIVES

Quality is a topic of interest throughout the present global market. As the market environment shrinks because of advancements in transportation and communication technologies, the need for rapid response to global developments becomes more critical. Price used to be the common basis on which product negotiations were consummated. In the new global market, product characteristics, expressed in quality terms, will be a major basis for such market negotiations. Manufacturers, retailers, distributors, vendors, and consumers will all need to have a common understanding of what constitutes acceptable product quality.

The striving for better quality worldwide has led to the need for unified international quality standards. The International Organization for Standardization (IOS) located in Geneva, Switzerland, has developed general quality guidelines known as ISO 9000. The IOS is a special international agency for standardization composed of the national standards organizations from several countries. The organization is also known as the International Standards Organization (ISO). The term ISO is more widely used than the term IOS.

The purpose of this book is not to present the specific details of the ISO 9000 standards themselves, but rather to present guidelines for successfully carrying out the ISO 9000 process. The documents containing the standards can be obtained directly from the standards bodies. Appendix A lists the potential sources of the standards documents.

GLOBAL COMPETITION

Political and economic changes now sweeping the world will significantly affect how and when products exchange hands in a globally competitive market. The quality of product will be a common basis for trade communication. Companies and countries must prepare for the following:

- Transition of some countries from being trade allies to being trade competitors.
- Reduction of production cycle time to keep up with the multilateral introduction of products around the world.
- Increased efforts to cope with the reduction in the life span of products.
- Increased responsiveness to the needs of a mixed workforce.
- Problems associated with overlapping cultural barriers.
- Increased need for multinational communication and cooperation.
- Need for multicompany and multiproduct coordination.
- Disappearance of trade boundaries.

DEFINITIONS OF QUALITY

Quality is often defined as a measure of customer satisfaction. Quality can be defined from the producer's point of view or the customer's point of view. However, to achieve its intended functions, a product must provide a balanced level of satisfaction to both the producer and the customer. For that purpose, a comprehensive definition of quality is required (Badiru and Ayeni 1993):

> *Quality refers to an equilibrium level of functionality possessed by a product or service based on the producer's capability and the customer's needs.*

Quality refers to the combination of characteristics of a product, process, or service that determines the product's ability to satisfy specific needs. The attainment of quality in a product is the responsibility of every employee in an organization. The production and preservation of quality should be a commitment that stretches from the producer to the customer. Products that are designed to have high quality cannot maintain the inherent quality at the user's end of the spectrum if the product is not used properly.

The functional usage of a product should match the functional specifications for the product within the prevailing usage environment. The ultimate judge of the quality of a product is the perception of the user. A product that is perceived as being of high quality for one purpose at a given time frame may not be seen as having acceptable quality for another purpose in another time frame. We can summarize quality rules as follows:

1. Quality is a product's ability to conform to specifications.
2. Specifications are a representation of customers' needs.
3. Quality begins at design.

THE NEED FOR STANDARDS

Standards provide a common basis for global commerce. Without standards, product compatibility, customer satisfaction, and production efficiency cannot be achieved. Just as quality cannot be achieved overnight, compliance with standards cannot be accomplished instantaneously. The process must be developed and incorporated into regular operating procedures over a period of time. Standards define the critical elements that must be taken into consideration to produce a high-quality product. Each organization must then develop the best strategy to address the elements.

Both *regulatory* and *consensus* standards must be taken into account when pursuing ISO 9000 requirements. Regulatory standards refer to standards that are imposed by a governing body, such as a government agency. All firms within the jurisdiction of the agency are required to comply with the prevailing regulatory standards. Consensus standards refer to a general and mutual agreement among a collection of companies to abide by a set of self-imposed standards. There may also be *contractual* standards that are imposed by the customer based on case-by-case or order-by-order needs. Most international standards will fall in the consensus category. Lack of international agreement often leads to trade barriers by nations, industries, and special interest groups.

INTERNATIONAL STANDARDS

Standards provide a basis for international conformance. Companies that expect to compete in the international market must comply with the prevailing standards. Many international standards have evolved in recent years as a direct response to new global market interactions. Although each country may have its own standards, there are worldwide standards that are uniformly applicable to all market participants. Difficulties in keeping up with the various national regulations and restrictions have led to calls for the standardization of standards.

Instead of having individual European standards, American standards,

4 IMPACT OF TRADE AGREEMENTS

FIGURE 1-1. ISO 9000 Uniting the World Market

Asian standards, and so on, we are beginning to see the emergence of global standards that are applicable to all countries. Such uniform standards are evolving from the prevailing high level of subscription to ISO 9000. But, contrary to popular belief, ISO 9000 itself is not a standard in the conventional sense. It is a mechanism through which a company can comply with standards. ISO 9000 provides a comprehensive review process with guidelines for how companies design, produce, install, inspect, package, and market their products.

Standards create avenues for identifying organizations that do not meet quality expectations and rewarding those that consistently meet them. Figure 1-1 illustrates the importance of ISO 9000 standards in world trade.

IMPACT OF TRADE AGREEMENTS

Trade agreements currently developing in different parts of the world will impact international quality standards. Agreements such as NAFTA (North American Free Trade Agreement) and GATT (General Agreement on Tariffs and Trade) will dictate how product quality issues will be handled in the international market. For example, GATT embodies a global system of trade rules. Companies and groups within and outside economic boundaries will be affected by changes in economic processes and industrial standards.

QUALITY AWARDS

Many quality awards have been established to encourage active participation in the quality movement. On a global level, the International Quality Europe Award, which highlights Europe's prevailing focus on quality, has been attracting applications from every corner of the world. The Standards Organization of Nigeria (SON) recently instituted the Nigerian Industrial Standards (NIS) certification and awards for local companies meeting international quality standards. At the national level, the Malcolm Baldrige National Quality Award has become the most coveted prize in U.S. industry. The award, started in 1988 by the U.S. Department of Commerce in memory of the late Commerce Secretary Malcolm Baldrige, is designed to honor companies that have shown the greatest commitment to quality improvement and management. The award is given in three categories: Manufacturing, Service, and Small Business. The premise of the award is to boost quality awareness in U.S. industries and prepare them to be more competitive in the international market.

The dramatic increase in interest in the Baldrige award is a clear indication that companies are becoming more and more quality conscious. Companies applying for the award spend hundreds of hours and thousands of dollars in preparing their applications. To emphasize results rather than mere commitment to quality improvement, applicants must demonstrate areas where they have been successful in achieving quality improvement. They must not only indicate the extent of their quality efforts, but must also prove that the efforts really work.

Many states in the United States have established their own state-level versions of the Baldrige award. For example, the Minnesota Quality Award (MQA), developed along the lines of the Baldrige award, has received national acclaim in the United States because of its integrative model. The MQA model consists of four major categories: *Driver, Systems, Measurement,* and *Goal.* The Driver is the driving force that facilitates, promotes, and serves as a champion of quality improvement efforts. Other state-level awards are the Massachusetts Quality Award and the Oklahoma State Quality Award. Numerous companies have also established company-wide awards that promote healthy competition between units of the same company.

The publicity, interests, and efforts generated by quality awards have significantly heightened the awareness of quality in many organizations. Many organizations that don't have immediate plans to apply for awards are already committing time and effort to quality improvement. The essential components of applying for any quality award include the following:

- Quality leadership
- Information and analysis
- Strategic planning for quality
- Human resources development and management
- Documentation of operational requirements
- Customer focus and satisfaction

PREPARING FOR THE CHANGE

Meeting the global competition requires change. An organization must be prepared for such necessary change. Quality improvement must be instituted in every aspect of everything that the organization does. If employees are better prepared for change, then positive changes can be achieved. The *"gain at great pain"* mode of quality improvement can be avoided if proper preparations have been made for changes. The requirements for getting an organization ready for the drastic changes that may be needed for quality improvement include the following:

- Keep everyone informed of the impending changes.
- Get all employees involved.
- Make employees a part of the decision-input mechanism.
- Highlight the benefits of improved quality.
- Train employees about their new job requirements.
- Create an environment for job enrichment.
- Allay fears about potential loss of jobs because of improved quality.
- Promote change as a transition to better things.
- Make changes in small increments.
- Make employees feel that they are owners of the change.

PLANNING LEVELS

As I have proclaimed before (Badiru 1993), "A plan is the map of the wise." A good plan for ISO 9000 is a company's roadmap to being more competitive in the market. Quality planning determines the nature of actions and responsibilities that are required to achieve a specified quality goal. Strategic quality planning involves the long-range aspects of quality improvement. Planning forms the basis for all actions. Strategic planning for change can be addressed at three distinct levels (Badiru 1991):

Supra-level planning: Planning at this level deals with the big picture of how quality improvement fits the overall and long-range organizational goals. Questions faced at this level may concern the potential contributions of quality improvement to the survival of the organization, the use of limited resources, the required interfaces with other projects within and outside the organization, the management support for the quality improvement, company culture, market share, customer expectations, and business stability.

Macro-level planning: Planning at this level addresses the overall planning within a defined product boundary. The scope of the improvement effort and its operational interfaces are addressed at macro-level planning. Questions faced at the macro level include goal definition, project scope, availability of qualified personnel, resource availability, project policies, communication interfaces, budget requirements, goal interactions, deadlines, and conflict resolution strategies.

Micro-level planning: This deals with detailed operational plans at the task levels of quality improvement. Definite and explicit tactics for accomplishing specific improvement objectives are developed at the micro level. Factors to be considered at the micro level of planning include scheduled time, training requirements, tools required, task procedures, reporting requirements, and quality-assurance steps.

TEAMWORK

Teamwork is essential to the success of any quality improvement endeavors. *Teamwork makes a team work.* Presented below are some of the desired characteristics of an effective ISO 9000 team:

- *A commitment to common goals*—ISO 9000 goals provide the team with a focus for operations. All members should have a clear understanding of the goals.
- *Mutual responsibility for goals*—It should be understood that goals can only be achieved through team effort.
- *Win-win approach*—Give a little to take a little. Compromise may be the name of the game in certain conflict situations. The team should jointly explore alternative ways in which organizational goals can be achieved without deleterious consequences to anyone.

- *Common agreement on high expectations*—All team members should strive to excel. Motivating ingredients for the team should be high standards, quality, and excellence. Mediocrity should not be tolerated.
- *Honest and open communications*—Members should openly and clearly express their thoughts and ask questions with confidence. There should be no hidden agendas.
- *Common access to information*—Information is a vital resource to the organization. It should thus be available and accessible to those who need it. It is the team leader's responsibility to ensure that information is available when needed.
- *Climate of trust*—Trust should be the bond that ties team members together. Technical and administrative respect within the team helps build the necessary trust.
- *Importance of each team member*—It should be recognized that each individual can have a significant impact on the quality of the final product. Thus, the opinion of each team member should be given the proper acknowledgment. Members should be assured that their ideas will be taken into consideration.
- *Support for team decision*—Each member should express the commitment to support overall team decisions. The team should be given assurance about the sincerity of the organization to support team decisions.

BENCHMARKING FOR ISO 9000

Metrics based on an organization's most critical quality issues should be developed. The objective of a company in international benchmarking is to match the best in the global market. *Benchmarking* is a process whereby target quality standards are established based on the best examples in the market. Benchmarking implies learning from the best. The premise of benchmarking is that, if an organization follows the best quality examples, it will become one of the best in the market. A major objective of benchmarking is to identify deficiencies in quality and attempt to remove them. The essential elements of benchmarking are as follows:

- Monitor prevailing standards in the market.
- Identify who is meeting the standards.
- Exchange information with market leaders.
- Determine/measure current performance level.
- Analyze gaps in performance.

Figure 1-2. Clearing ISO 9000 Hurdles with Management Leadership

- Continuously strive to close the gaps.

Once the benchmarks have been set, a company must seize the initiative and create new market leadership. The company must

1. Go beyond the benchmarks. View the benchmarks as the minimum achievement levels.
2. Be innovative and develop unique product or process concepts. This creates a proprietary opportunity that will keep the company ahead of competitors.
3. Play a highly visible and audible role in the market. The company must participate in any publicity opportunities. Whether you call them "publicity stunts" or "advertising gimmicks," they will keep the company's name in the mind of the market.
4. Expand market territory. Marketing and sales activities must be intensified to keep pace with the global market movements. Focus on the long-range goal of the company.
5. Liaise with the global customer. Create access to both direct and indirect customers across the traditional "imaginary" market boundaries. Improved political relations and better communication technologies now available around the world should be used to advantage to reach out to the customers.

MANAGEMENT LEADERSHIP

A fish always starts to rot at the head. So says a Turkish proverb. Quality problems often start at the top. Management leadership will invariably determine the level of quality in an organization. Without strong management leadership, the objectives of ISO 9000 cannot be achieved. Although ISO 9000 is everybody's responsibility, the quality of management leadership will determine its success or failure. Management should take the lead in paving the way to clear the ISO 9000 hurdles as depicted in Figure 1-2.

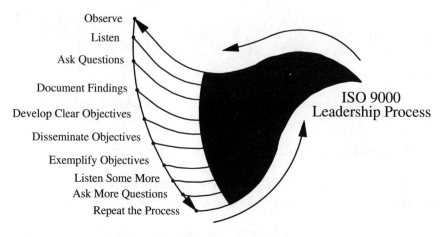

Figure 1-3. ISO 9000 Leadership Process

Traditional managers manage workers to work, so there is no point of convergence or active participation. Modern managers team up with workers to get the job done. Figure 1-3 presents a practical leadership model of ISO 9000. The model embodies the following crucial elements: observe, listen, ask questions, document findings, develop clear objectives, disseminate the objectives, participate in an exemplifying role, listen some more, ask more questions, and repeat the process.

Good leadership is an essential component of project management. Project leadership involves dealing with managers and supporting personnel across the functional lines of the project. It is a misconception to think that a leader leads only his or her own subordinates. Leadership responsibilities can cover functions vertically up or down. A good project leader can lead not only his or her subordinates, but also the entire project organization including the highest superiors. Leadership involves recognizing an opportunity to make improvement in a project and taking the initiative to implement the improvement. In addition to inherent personal qualities, leadership style can be influenced by training, experience, and dedication. The following section outlines the categories, roles, and responsibilities of the various leadership groups involved in the pursuit of ISO 9000 certification.

Upper Management Responsibilities

Management responsibilities for ISO 9000 registration process can be summarized into the following essential elements:

1. Designate a management representative to champion the cause of

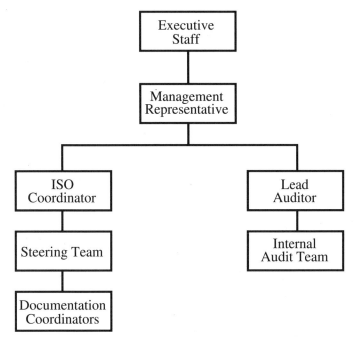

Figure 1-4. Model for ISO Organizational Structure

ISO 9000. Establish a steering committee to liaise with management. Figure 1-4 presents a simple model of an organizational structure for ISO 9000.
2. Be familiar with the general aspects of ISO standards and be sure that someone in the organization is very conversant with the details. The ISO representative should know what sections apply to specific functional areas.
3. Ensure that work groups comply with needs and changes required by ISO 9000 as determined by steering committee.
4. Set an ultimate goal of expanding ISO standards compliance throughout the organization.
5. Provide the time and resources required, within reasonable business constraints, for ISO preparation, audit, and follow-up.

Role of Management Representative

- Obtain resources for closed-loop corrective actions.
- Work with appropriate management to resolve nonconformance problems.

- Propose improvements.
- Serve as primary contact for the certifying agency.
- Serve as focal point for third-party audits.
- Resolve conflicts or problems encountered during third-party audits.

Role of ISO Coordinator

- Work under the direction of the management representative.
- Form a steering team.
- Schedule meetings, prepare agenda, facilitate meetings, obtain resources.
- Plan and organize implementation of conformance requirements.
- Communicate progress to local and corporate management.
- Provide communications to the entire organization concerning ISO process.
- Liaise with other internal ISO coordinators.
- Benchmark with other businesses and companies.

Role of Lead Auditor

- Serve as the audit leader.
- Plan, implement, and maintain the auditing system.
- Ensure coverage of all aspects needed to achieve certification.
- Develop a management policy that facilitates the ISO process.
- Coordinator manpower, funding, and facilities for audit system.
- Form audit team as needed; select qualified, responsible people.
- Coordinate training.
- Provide reasonable and timely access to required documents.
- Report on the effectiveness of the quality assurance system.
- Liaise with personnel to ensure corrective action.

Role of Internal Auditor

- Maintain education, technical ability, and temperament to advise or assist on a professional basis in identifying, defining, and solving specific quality system problems involving the organization, planning, direction, and control of the activity.
- Serve the audited organization as an impartial, objective observer following a certain set of quality objectives as defined by policies and procedures of the organization.

- Exhibit an outstanding ability to think and solve problems (analytical, diagnostic, and synthesizing ability).
- Exhibit high-level skills in oral and written communications.

Role of Documentation Coordinator

- Work directly with organization executive to comply with ISO requirements.
- Ensure that organizational policies and procedures are up-to-date.
- Maintain departmental policy and procedure files as required.
- Serve as focal point for internal and third-party audits.
- Serve as contact for ISO steering committee.

RISK MANAGEMENT

Quality begins with design, but should not end there. As the traditional producer's "say-so" slowly transforms into the buyer's discretion, it is essential to harmonize the risks of producers and consumers in the evaluation of product quality. Both must be aware of their respective rights and responsibilities in the market. For the producer, designing quality into the product will lessen the risk to both the producer and the consumer. However, it is unnecessarily expensive and counterproductive to over-design a product just because higher quality is desired. For the consumer, it is irrational to demand the highest possible quality if the operating environment cannot accommodate such quality level. The risks must be weighed against the benefits. Compliance with ISO 9000 standards can reduce the risk of product liability.

There is a story of a buyer who was promised a robust design by the producer. The buyer interprets "robust" to mean a big, strong product. When the product was finally delivered, the buyer was not satisfied because the product appeared to be a miniature representation of the "robust" product that he had expected. What the producer actually promised was what was delivered: a *versatile,* not necessarily big, product. This is a case of buyer-producer miscommunication.

In statistical parlance, the producer's risk is referred to as *Type I error* and the consumer's risk is referred to as *Type II error.* For example, we may have a batch of products from a production process and we want to evaluate the relative risks for the producer and the consumer. Rejecting the batch when it is good implies a risk to the producer, because a rejected batch never makes it to the market. On the other hand, accepting the batch when it is bad implies a risk to the consumer, because there is the potential that a consumer will end up with a bad product. Table 1-1 illustrates the conventional view that puts products in mutually exclusive quality categories.

TABLE 1-1 Producer's Risk versus Consumer's Risk

	Good Batch	Bad Batch
Accept Batch	Correct decision $1 - \alpha$	Type II Error (β: Consumer's Risk)
Reject Batch	Type I Error (α: Producer's Risk)	Correct decision $(1 - \beta)$

In the table, α refers to the probability of Type I error. That is, the probability of rejecting a product when it is actually good. The letter β refers to the probability of Type II error, that is, the probability of accepting a product when it is actually bad. In actual practice, there is a gray line between product rejection and acceptance. This gray line may put a product in either category based on suitability for certain types of functions.

Thus, a product that is rejected for one functionality may be acceptable for another functionality. This is illustrated in the expanded risk categories shown in Table 1-2. Note that a bad product that is put to alternate use may constitute a high risk to the user. A marginal product that is accepted represents a medium risk to the user. A marginal product that is rejected represents a medium risk to the producer. A marginal product that is put to alternate use represents a low risk to either the producer or the consumer. This low risk may be in terms of lower market price for the product and reduced functionality to the consumer.

ISO 9000 MISSION LOGISTICS

The road to ISO 9000 certification may be filled with obstacles. But the obstacles can be overcome with proper awareness, planning, and a logistical approach. Figure 1-5 illustrates an ISO 9000 Mission Logistics. Every orga-

TABLE 1-2 Expanded Quality Risk Categories

	Good Product	Marginal Product	Bad Product
Accept	Correct decision $1 - \alpha$	Medium Risk	Type II Error (β: Consumer's Risk)
Alternate Use	Medium Risk	Low Risk	High Risk
Reject	Type I Error (α: Producer's Risk)	Medium Risk	Correct decision $(1 - \beta)$

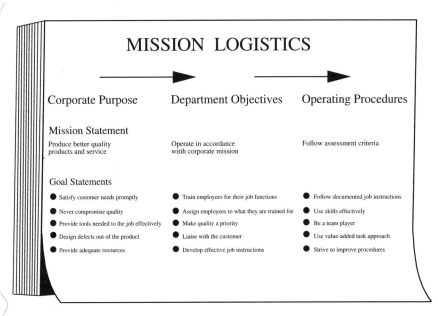

Figure 1-5. ISO 9000 Mission Logistics

nization's ISO 9000 logistics should cover a mission statement, goals, expected outcomes, assessment criteria, and procedures.

According to the Mission Logistics, no individual should operate in a vacuum. Each employee must operate within a departmental structure. The department must operate within the corporate structure. The corporate mission statement propels the department mission statement, which should propel individual operating procedures. Employees and management must exhibit mutual responsibility for quality. With this hierarchical relationship, everyone would be working towards a common goal—quality improvement.

2
INTRODUCTION TO ISO 9000

Many international bodies have been either newly organized or expanded to cater to the emerging quality issues. The emergence of ISO 9000 is one of the results of heightened awareness about quality and standardization. ISO 9000 reflects the current global trend toward ever-increasing stringent consumer expectations with respect to quality.

QUALITY PROCLAMATION

Giving the right signals is crucial to securing support, cooperation, and commitment for ISO 9000 efforts. Top management and senior managers must institute a company-wide proclamation about quality. Figure 2-1 presents the author's concept of what a company's quality proclamation may contain. Such company-wide proclamation is essential to motivate everyone about quality improvement and the requirements of ISO 9000.

WHAT IS ISO 9000?

ISO 9000 is a set of five individual but related international standards on quality management and quality assurance. The standards were developed to help companies effectively document the quality system elements required to maintain an efficient quality system. ISO 9000 standards set the platform for global competition.

Figure 2-1. Company's Quality Proclamation

The standards were introduced in 1987. They are not specific to any particular industry, product, or service. In the United States, the standards are referred to as **ANSI/ASQC Q90 series.** The ANSI/ASQC Q90 through Q94 standards are technically equivalent to the ISO 9000-9004 series. The American National Standards Institute (ANSI) is the member body representing the United States in the International Standards Organization (ISO). The standards are implemented as **EN 29000** within the European Economic Community (EEC) and as **BS 5750** in the United Kingdom. ISO has about 180 technical committees responsible for specific areas of specialization. The objective of ISO is to promote the development of standardization and cooperation in the international exchange of goods and services.

ISO 9000 is not a technical product. It is not hardware. It is not software. It is not a mere concept. A company can't just go out and buy it. It is a quality improvement process that must be cultivated in-house. A key strategy is not to limit ISO registration to only the manufacturing activities. Distribution, design, preproduction, production, subassembly, and component facilities should all be candidates for ISO 9000 certification. ISO 9000 certification certifies that an organization is doing what it says it is doing with respect to operating procedures. Even for individuals, following ISO 9000 guidelines can create the discipline needed to achieve better performance in any endeavor, be it personal, public, or professional. Many organizations now voluntarily comply with ISO 9000 requirements even though they may not necessarily be pursuing formal certification.

Benefits of ISO 9000

- Establishes consistent quality practices that cross international borders.
- Provides common language and terminology.
- Provides a reference point for trade negotiations.
- Minimizes the need for on-site customer visits or audits.

Benefits of Compliance

- Creates internal value and pride in organizational process.
- Facilitates business survival.
- Creates opportunity for market expansion.
- Reduces the risk of product liability.
- Useful as a marketing tool.
- Leverage for new markets, products, or industries.
- Demonstrates commitment to quality.
- Establishes proactive response to international market demands.
- Complements TQM concepts.
- Builds customer confidence.
- Formalizes the quality system with an organization.

WHAT'S IN THE NAME?

The short designation for ISO 9000 was borrowed (as a pun) from the Greek word *isos,* which means "equal." *Isos* is used as the root for many words having to do with equality. For example, *isometric* refers to equal measures or dimensions; *isonomy* refers to equality of people in the eye of the law; *isothermal* refers to the presence of equal temperatures; and *isotropic* refers to invariance with respect to direction. *ISO 9000* is intended to convey the idea of the invariance that is possible when a standard is available. When a standard is available for a process, the process is expected to produce identical or invariant units of a product. That is, *iso-product* or *iso-units.* In an attempt to justify the name ISO 9000, people generally misinterpret the organization's name, IOS, as ISO. There is only a cursory correlation between the ISO in the organization acronym and the ISO in the name of the standards.

ISO 9000 registration confirms whether or not a company has conscientious quality procedures. ISO 9000 iş the master plan for quality improvement. Approaches such as total quality management (TQM) and continuous process improvement (CPI) are the specific maps for negotiating the plan.

What about Terminology?

An organization should not get hung up on terms used in the ISO standards documents. The intent of the requirement is more important than the words being used. Terminology changes from one industry to another, and the ISO standard is worded to embrace many industries and services. An important issue is that ISO 9000 depends on the quality system as actually implemented, not just as documented.

SUBDIVISIONS OF ISO 9000

ISO 9000 actually consists of five subdivisions. All the subdivisions contain a set of models and guidelines for quality assurance and quality management. The standards can be applied in flexible modes. The approach will vary from company to company based on specific needs. However, the end result must please external ISO auditors. The following explains each subdivision of ISO 9000.

ISO 9000—*Quality Management and Quality Assurance Standards: Guidelines for Use.* This is the roadmap that provides guidelines for selecting and using 9001, 9002, 9003, and 9004. A supplementary publication, ISO 8402, provides quality-related definitions. ISO 9000 and ISO 9004 are for guidance in the use of the standards.

ISO 9001—*Quality Systems: Model for Quality Assurance in Design/ Development, Production, Installation and Servicing.* This is the most comprehensive standard. It contains 20 elements and presents a model for quality assurance for firms involved in the design, manufacturing, and installation of products and/or services.

ISO 9002—*Quality Systems: Model for Quality Assurance in Production and Installation.* This contains 18 elements and is for firms involved in manufacturing or production of products and/or services only. The requisite design is usually specified by customers.

ISO 9003—*Quality Systems: Model for Quality Assurance in Final Inspection and Test.* This contains 12 elements and is for firms involved in the distribution, inspection, and testing of manufactured products or services only, without any production or installation activities. It presents a model for quality assurance in final inspection and test.

ISO 9004—*Quality Management and Quality System Elements: Guidelines.* This provides guidance for a supplier to use in developing and implementing a quality system and to determine the extent to which each quality system element is applicable. It examines each of the quality system elements in greater detail and can be used for internal and external auditing purposes. It contains guidelines for users in the process of developing in-house quality systems.

The ISO 9000 standards require a written documentation of every aspect of a business process so that all employees are aware of and consistently comply with written work procedures. The ISO 9000 team in an organization must keep the various ISO standards in perspective. ISO 9004 presents the elements of a complete quality management system. ISO 9001 presents the minimum requirements to become certified. The focus of effort should be on ISO 9001, using ISO 9004 as it is intended, as a guide.

PURPOSE OF ISO 9000

Conventional standards are often applied to specific products or operations. To the contrary, ISO 9000 is applied throughout an organization. The standards can be applied to both manufacturing and service functions. Even individuals can apply the standards to their own specific functions. Compliance with the standards helps increase acceptability in international markets as well as improving internal productivity and performance.

In addition, the standards help in determining capable suppliers with effective quality assurance systems. The standards help reduce buyers' quality costs through confidence and assurance in suppliers' quality practices. Compliance with an ISO 9000 standard provides a means for contractual agreement between the buyer and the supplier. Companies that are certified and registered as meeting the ISO standards will be perceived as viable suppliers to their customers. Those that are not will be perceived as providing less desirable products and services.

The standards are designed to address a variety of quality management scenarios. For example, if a supplier has only a manufacturing facility with no design or development function, then ISO 9002 would be used to evaluate the quality system. Each country has its own quality system standards that relate to the ISO 9000 standards. Each individual company is encouraged to formally register for compliance with the standards. In fact, a request for a supplier's ISO 9000 registration number has become an important element when companies make their selection of suppliers.

The ISO 9000 series standards define the minimum requirements a supplier must meet to assure its customers that they are receiving high-quality products. This has had a major impact on companies around the world. Through the ISO standards, suppliers can now be evaluated consistently and uniformly. The ISO 9000 series has been adopted in the United States by the American National Standards Institute (ANSI) and the American Society for Quality Control (ASQC) as ANSI/ASQC Q90 standards. The European equivalent of ISO 9000, named EN 290000 series, is also widely recognized in world markets. Table 2-1 presents a taxonomy of some of the major quality assurance standards around the world. Most of the standards are technically equivalent except that they incorporate specific and unique considerations of the country involved.

WHAT IS THE QUALITY SYSTEM?

For ISO 9000 certification, what is registered is the *quality system* used to produce a product, not the product itself. Registration of the quality system does not necessarily imply product conformity. ISO 9000 registration is quickly becoming a crucial part of commercial activities. It is believed that within a few years, ISO 9000 registration will be necessary to stay in business and be competitive.

Like any conventional system, a quality system is a collection of interrelated elements with a common objective. The usual elements of a quality system are

- Human resources
- Capital
- Equipment
- Skills
- Methods
- Policies
- Procedures
- Other resources

All of these elements together constitute the synergistic system needed to achieve the objective of quality improvement. Essentially, the quality system consists of the organization, structure, resources, responsibilities, procedures, and processes that are used to manage quality. They must be documented so that they are understood by the appropriate people and maintained at a level that facilitates consistent control.

TABLE 2-1 Quality Assurance Standards

Standards body or country	Quality Management and Quality Assurance Standards: Guidelines for Selection and Use	Quality Systems Model for Quality Assurance in Design, Development, Production, Installation and Servicing	Quality Systems Model for Quality Assurance in Production and Installation	Quality Systems Model for Quality Assurance in Final Inspection and Test	Quality Management and Quality System Elements: Guidelines
ISO	ISO 9000: 1987	ISO 9001: 1987	ISO 9002: 1987	ISO 9003: 1987	ISO 9004: 1987
Australia	AS 3900	AS 3901	AS 3902	AS 3903	AS 3904
Austria	OE NORM-PREN 29000	OE NORM-PREN 29001	OE NORM-PREN 29002	OE NORM-PREN 29003	OE NORM-PREN 29004
Belgium	NBN X 50-002-1	NBN X 50-003	NBN X 50-004	NBN X 50-005	NBN X 50-002-2
Canada	—	—	—	—	CSA Q420-87
China	GB/T 10300.1-88	GB/T 10300.2-88	GB/T 10300.3-88	GB/T 10300.4-88	GB/T 10300.5-88
Denmark	DS/EN 29000	DS/EN 29001	DS/EN 29002	DS/EN 29003	DS/EN 29004
Finland	SFS-ISO 9000	SFS-ISO 9001	SFS-ISO 9002	SFS-ISO 9003	SFS-ISO 9004
France	NFX 50-121	NFX 50-131	NFX 50-132	NFX 50-122	NFX 50-133
Germany	DIN ISO 9000	DIN ISO 9001	DIN ISO 9002	DIN ISO 9003	DIN ISO 9004
Hungary	NI 18990: 1988	NI 18991: 1988	NI 18992: 1988	NI 18993: 1988	NI 18994: 1988
India	IS: 10201 Part 2	IS: 10201 Part 4	IS: 10201 Part 5	IS: 10201 Part 6	IS: 10201 Part 3
Ireland	IS 300 Part 0	IS 300 Part 1	IS 300 Part 2	IS 300 Part 3	IS 300 Part 4
Italy	UNI/EN 29000-1987	UNI/EN 29001-1987	UNI/EN 29002-1987	UNI/EN 29003-1987	UNI/EN 29004-1987
Malaysia	—	MS 985/ISO9001: 1987	MS 985/ISO9002: 1987	MS 985/ISO9003: 1987	—
Netherlands	NEN-ISO 9000	NEN-ISO 9001	NEN-ISO 9002	NEN-ISO 9003	NEN-ISO 9004
New Zealand	NZS 5600-1987 (Part 1)	NZS 5601-1987	NZS 5602-1987	NZS 5603-1987	NZS 5600-1987 (Part 2)

(continued)

TABLE 2-1 Continued

Nigeria	NIS/ISO 9000	NIS/ISO 9001	NIS/ISO 9002	NIS/ISO 9003	NIS/ISO 9004
Norway	NS-EN 29000: 1988	NS-EN 29001: 1988	NS ISO 9002	NS ISO 9003	—
Russia	—	40.9001- 88	40.9002- 88	—	—
South Africa	SABS 0157: Part 0	SABS 0157: Part I	SABS 0157: Part II	SABS 0157: Part III	SABS 0157: Part IV
Spain	UNE 66 900	UNE 66 901	UNE 66 902	UNE 66 903	UNE 66 904
Sweden	SS ISO 9000: 1988	SS ISO 9001: 1988	SS ISO 9002: 1988	SS ISO 9003: 1988	SS ISO 9004: 1988
Switzerland	SH ISO 9000	SH ISO 9001	SH ISO 9002	SH ISO 9003	SH ISO 9004
Tunisia	NT 110.18- 1987	NT 110.19- 1987	NT 110.20- 1987	NT 110.21- 1987	NT 110.22- 1987
United Kingdom	BS 5750: 1987, Part 0 (Section 0.1) ISO 9000/EN 29000	BS 5750: 1987, Part 1 ISO 9001/EN 29001	BS 5750: 1987, Part 2 ISO 9002/EN 29002	BS 5750: 1987, Part 3 ISO 9003/EN 29003	BS 5750: 1987, Part 0 ISO 9004/EN 29004
USA	ANSI/ASQC Q90	ANSI/ASQC Q91	ANSI/ASQC Q92	ANSI/ASQC Q93	ANSI/ASQC Q94
Yugoslavia	JUS A.K. 1.010	JUS A.K. 1.012	JUS A.K. 1.013	JUS A.K. 1.014	JUS A.K. 1.011
EEC	EN 29000	EN 29001	EN 29002	EN 29003	EN 29004

As is the classical definition of a system, a quality system has some unique characteristics that an organization should understand and promote. These characteristics are

1. Interaction with the environment
2. Possession of an objective
3. Self-regulation capability
4. Self-adjustment capability

With respect to quality management, interaction with the environment may be defined in terms of what the market environment (the customer) wants. The objective of the quality management system is to achieve an acceptable level of quality. The self-regulation characteristic relates to the system's ability to maintain the stipulated quality level once it is achieved. The self-adjustment characteristic relates to the system's ability to make amendments should the quality level deviate significantly from the required level. The acceptable quality level itself should not be stagnant. It should be revised periodically and upgraded as market needs change. The major requirements of a complete quality system follow. Chapter 4 presents detailed explanations of each.

1. Management responsibility
2. Quality system documentation
3. Contract review
4. Design control
5. Document control
6. Purchasing
7. Purchaser supplied product
8. Product identification and traceability
9. Process control
10. Inspection and testing
11. Inspection, measuring, and test equipment
12. Inspection and test status
13. Control of nonconforming product
14. Corrective action
15. Handling, storage, packaging, and delivery
16. Quality records audits
17. Internal quality audits
18. Training
19. Servicing
20. Statistical techniques

In all of the above requirements, the corporate culture of the organization will determine the level of success that can be achieved. Two important axioms that an organization should abide by are

- Treat every customer as if he or she is the last customer you will ever get.
- Treat every employee as if he or she is the last help you will ever get.

ACCEPTANCE OF ISO 9000

Many companies assert that ISO registration would influence their choice of suppliers. Many have also reported significant cost savings in their operations as a result of pursuing ISO registration. To be competitive, suppliers are also voluntarily pursuing registration. ISO registration has been slow in the United States. As of this writing, there are fewer than two thousand ISO-certified U.S. manufacturers. This is a low tally compared to the over 15,000 registrations in Great Britain. The registration pace has been very slow despite the fact that ISO 9000 is now widely accepted in many companies, but it is expected to accelerate as more easy-to-follow implementation guidelines become available.

It is recognized that the pursuit of ISO 9000 standards may be relatively too expensive for small companies, particularly those in job-shop environments. However, the culture of ISO 9000, whether certification is formally sought or not, should benefit even the smallest companies. The discipline introduced by ISO 9000 will enhance the operating characteristics of any organization.

Benefits of ISO 9000 Registration

Corporate Reasons for ISO 9000 Registration

- Customer demands and expectations
- Internal benefits of higher quality
- Opportunity for market advantage
- Compliance with corporate requirements
- Competitive pressures
- Reduced costs of operation
- Market prestige
- Mandatory requirement for export

Internal Benefits

- Better documentation of process
- Greater awareness of quality
- Better employee morale
- Higher operational efficiency and productivity
- Increased process coordination
- Better communication and cooperation
- Reduced scrap
- Fewer lost-time accidents

Market Benefits

- Worldwide acceptance of ISO 9000

- Company prestige
- Better customer satisfaction
- Higher customer trust
- Improved competitive edge
- Reduced need for customer audits
- Increased market share
- Quicker responsiveness to market needs

Concerns About ISO 9000

- ISO 9000 consultants are not regulated
- Lack of acceptance of local auditors abroad
- No effective centralized record of registrations
- High cost of certification for small companies
- Limited understanding of the process

The ISO standards carry an often overlooked contractual obligation. The customer may specify in a contract that the supplier's quality system shall meet specific ISO 9000 requirements. In such a case, the quality system becomes a binding contractual requirement. Therefore, changes to the quality system must be coordinated with the customer prior to implementation.

It is important to realize that there are only a few number of products for which ISO 9000 is a *mandatory* requirement. These are mainly in

- Products in the medical devices area
- Sensitive safety products in the personal protective equipment area, such as respirators used in poisonous atmospheres
- Other areas of product safety, where is there is a product breakdown, there will be catastrophic effect on people using the product

GETTING STARTED

There are two aspects of awareness relating to ISO 9000. This book provides guidelines for declarative knowledge and procedural knowledge. Declarative knowledge relates to knowing what the ISO 9000 requirements are. Procedural knowledge relates to the procedures for complying with ISO 9000 requirements. The procedures include easy-to-follow steps and examples of implementing ISO 9000 process. In simple terms, declarative knowledge shows what needs to be done, while procedural knowledge shows how to do it. An effective approach to disseminating ISO 9000 knowledge is embodied in the following pointer: Use the Triple C Model for implementing ISO 9000. The Triple C model for *communication, cooperation,* and *coordination* is described in detail in Chapter 6.

It was estimated that the first-time failure rate for American companies seeking ISO 9000 registration in 1993 was as high as 50 percent. This shows a serious lack of preparation. ISO 9000 tells companies what they have to do, but not how to do it. This book presents guidelines for how to do what needs to be done. The guidelines can help companies be properly prepared for the registration requirements.

Major Phases of ISO 9000 Registration Process

Planning and Preparation Phase

1. Decision to comply
2. Education and training
3. Determination of scope of compliance

Compliance Phase

1. Assess current system
2. Address gaps
3. Document existing quality system
4. Establish internal audit function

Registration Phase

1. Decision for third-party registration
2. Select a registrar
3. Third-party assessment
4. Address gaps
5. Ongoing maintenance of registration (third-party surveillance audits)

Not every ISO 9000 requirement will be applicable or needed by every organization. An ISO 9000 responsibility matrix (see Chapter 6) can be used to show each organization any ISO requirements for which the organization is responsible. The ISO coordinator, management representative, steering committee, and internal audit team should be the major players in implementing the ISO 9000 process. Even companies that have been "ISOed" should continue to strive for compliance. Compliance with ISO 9001 involves the following:

1. Adequately defining the work processes and relationship
2. Sufficiently documenting the processes
3. Executing the processes in accordance with the documented procedures
4. Recording, tracking, and verifying corrective actions
5. Periodically auditing the processes to assure compliance and continuous quality improvement

INTERNAL AUDIT

The philosophy of quality assurance is to work toward prevention rather than detection of problems. Internal auditing is useful in identifying potential problem areas so that corrective actions can be taken before problems actually develop. It examines the effectiveness of management in implementing control programs. Some of the specific purposes for internal auditing include:

- Ensuring that written procedures exist, are adequate, and are followed,
- Verifying adherence to legal or regulatory requirements,
- Identification of deficiencies in the management system,
- Confirming that corrective actions are effective and are carried out, and
- Providing management with objective evidence based on facts.

In addition to helping prevent problems, internal auditing can also help when problems do occur. In this case emphasis is placed on

- Early detection of the problem,
- Assessment of the extent and depth of the problem, and
- Identification of the root cause of the problem.

Checklist

A checklist can be very useful when performing an internal audit. Checklists can be used for a preliminary survey of specific areas for improvement such as *employees, materials, layout,* and *job design.* The ambient work atmosphere can significantly affect the quality of work.

Employees

☐ Are employees properly trained?
☐ Are working hours effectively used?

- ☐ Is the workforce stable?
- ☐ Are employees happy and motivated?
- ☐ Does the company have excessive absenteeism?
- ☐ Are employees punctual?
- ☐ Do employees produce acceptable quality of work most of the time?
- ☐ Do employees know their next job assignments?
- ☐ Is work scheduled effectively?
- ☐ Are documented methods used for better training?
- ☐ Are recommended methods enforced?
- ☐ Are there time standards? Are employees aware of them?

Materials

- ☐ Can quality specifications be met with materials provided?
- ☐ Are established procedures followed?
- ☐ Are material requirements realistic?
- ☐ Are supplies provided in the quantity needed?
- ☐ Do employees understand the cost of material waste?
- ☐ Are proper tools and equipment provided?

Layout

- ☐ Does layout facilitate high-quality work?
- ☐ Is material handling excessive?
- ☐ Is work performed in the proper sequence?
- ☐ Is aisle space adequate for the type of work?
- ☐ Are supplies stored effectively?
- ☐ Are tools conveniently located?
- ☐ Is lighting adequate?
- ☐ Is there good general housekeeping?

Job Design

- ☐ Is the job well-documented?
- ☐ Is the job necessary?
- ☐ Is technical information available at the job site?
- ☐ Can the value added by the job be identified?
- ☐ Can a simpler operation be used?
- ☐ Can a more economical process be used?
- ☐ Can the job be combined with other jobs?

- ☐ Is the simplest method employed?
- ☐ Does the job frequently need expediting?
- ☐ Does the job design complicate the job?
- ☐ Does the job require excessive setup?
- ☐ Does excessive paperwork impede the job?

THIRD-PARTY AUDIT

In order to become registered, a company must first complete the certification process. This is initiated by selecting an independent third-party assessor to come to the company's facilities and evaluate the systems that are used for designing, producing, calibrating, revising, testing, marketing, and shipping products. The third party, usually a local standards organization (e.g. Underwriters Laboratory [UL]), acts as an independent body in evaluating a supplier's quality system. In the United States, the American Registrar Accreditation Board oversees the activities of ISO registrars.

It is required that documentation and records be available and current. The third-party assessor is known as a registrar and should be selected in the same manner as any subcontractor. For initial assessment, the registrar will send an audit team to the company for an extensive assessment of each business and function. A report is prepared to list the nonconforming areas.

A corrective action plan would then be developed for each area of deficiency and submitted to the registrar for approval. The registrar will continue with follow-up assessments at the rate of about two per year until the entire ISO 9000 standards have been covered. The registrar ensures that the quality system has been defined, documented, and implemented, and that it satisfies the requirements of the appropriate standard. The third-party audit is an appraisal of the quality management system, and not an assessment of specific product quality. The objective is to verify the design, implementation, and effectiveness of an organization's quality system.

If the company's quality system conforms to the registrar's interpretation of an ISO 9000 series standard, the company is given a *Certificate of Registration* (is certified) for one of the ISO 9000 quality assurance standards, ISO 9001, 9002, or 9003. The certification is then listed in a general registrar, which is available to the public. Certification allows the company to use the *Registered Firm Symbol* provided by the registrar on advertising, correspondence, invoices, and so on, as evidence of ISO registration. The number shown underneath the Registered Firm Symbol refers to the Certificate of Registration issued to the certified company. Some of the key elements that the third-party auditors check for in a company are as follows:

- Does the company have a documentation process? Does the documentation provide adequate guidelines for workers?
- Is everyone in the company following the documented process? Is everyone aware of updates and changes to the documentation?
- How are materials selected? Are appropriate materials selected for specific processes?
- How are in-house inspections of suppliers' deliveries handled? Is the company getting what it wants from suppliers?
- Calibration and metrology processes. Are calibrations done properly? Are measurements being made accurately?
- The procedure for taking corrective actions. Are avenues available for identifying, reporting, and correcting problems?
- The internal self-auditing process. Are problems overlooked when they are identified? Is there a formal process-review policy? Is the company defensive about obvious quality problems?

A successful ISO 9000 audit is a prerequisite for ISO 9000 registration. Registration affirms that a company is meeting acceptable quality standards. Even after registration, the auditors come back periodically to make sure that the standards continue to be met. Good documentation helps each employee to know exactly what is expected of him or her with respect to the quality of products and services. This awareness can affect morale positively and provide the impetus for further personal commitment to quality.

An audit determines if (1) a complete and documented quality system is in place, (2) the system is fully functional, (3) the system is self-correcting, (4) the documentation and contents comply with standards, and (5) the laid-down procedures are being followed in actual practice.

Third-Party Assessment Steps

Step 1: Initial Contact
   Contact a certification body
Step 2: Information Exchange
   Submit quality manual and other requested information
Step 3: Preliminary Site Visit
Step 4: Arrival Conference
   Explain purpose and scope
   Review audit schedule

  Discuss other details
Step 5: Conduct Audit
 Express full cooperation
 Review of objective evidence
  • *Interview management and workers*
  • *Examine documents and records*
  • *Discuss observations*
 Document findings
 Optional feedback sessions
Step 6: Exit Conference
 Review process, questions, etc.
Step 7: Formal Report
Step 8: Follow-up Audit (If Required)
Step 9: Surveillance Audit
 Every six months
 Unannounced audit visits

After a company has achieved ISO 9000 certification, the registrar monitors the quality system of the company periodically and withdraws the registration when conformity with the standard is not effectively maintained.

DOCUMENTATION STEPS

Throughout the ISO standards, the two key elements that are stressed again and again are *documentation* and *documentation control.* Documentation should be an integral part of the compliance strategy of the organization. Documents that directly affect the acquisition of materials, design of the product and processes, organizational work processes, and responsibilities should be carefully developed and maintained.

  The common documentation deficiency in many companies involves documentation of organizational work processes and functional responsibilities. Another frequent documentation problem in industry is outdated documents. To reflect current organizational process, documentations should be reviewed for content and currency. When appropriate, changes must be made to reflect current organizational alignments and associated responsibilities. These areas of deficiency are where project management techniques can be highly effective.

  Figure 2-2 presents an effective hierarchy of documentation for ISO 9000. The elements of documentation are defined as follows:

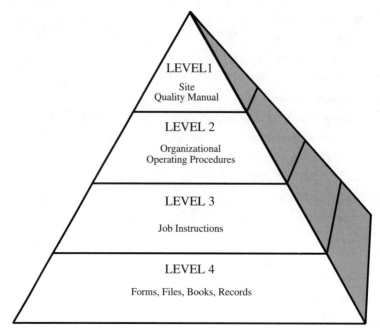

Figure 2-2. ISO 9000 Document Hierarchy

Quality system manual gives an overview of the quality management system. It is arranged in order of the appropriate ISO standard.

Standard operating procedure defines cross-functional activities or major segments of the system.

Functional procedure describes an organization in terms of mission, inputs, outputs, and flow or value-adding steps.

Work instructions are specific instructions unique to a particular operation or organization. They are controlled by the document center when they describe processes that directly affect product—process planning, drawings, and specifications. Work instructions in engineering and support organizations are usually controlled by the individual departments.

Quality records are documents, files, forms, or data that show achievement of required quality or show effective operation of the quality system.

Documentation Process Summary

The documentation process can be summarized in five steps:

Step 1: Understand External Requirements
- *ISO requirements*

- *Industry/regulatory requirements*
- *Corporate requirements*

Step 2: Understand Your Business
- *Functional organizational chart*
- *Mission*
- *Inputs of each major segment*
- *Outputs of each segment to the rest of the organization*
- *Flow chart*

Step 3: Assess Current Documentation System
- *Responsibilities for ownership, creation, approval*
- *Document storage, retrieval, control*
- *Survey existing documents and quality records*
- *Gap analysis; benchmarking*

Step 4: Develop a Documentation Structure
- *Define each level of documentation*
- *Define functional interfaces*
- *Establish a document format*
- *Create a document control system*
- *Create a quality record format and storage criteria*

Step 5: Create Individual Documents
- *Establish responsibilities, priorities*
- *Create operating documents*
- *Establish quality records*
- *Write the quality manual*

Documentation can be achieved in four distinct levels to facilitate effectiveness. The levels are outlined below.

Documentation Level 1: Quality Manual (site-specific)

- Organized to ISO 9001
- General description (20 to 40 pages)
- Serves as the controlling document
- Serves as a marketing tool

Documentation Level 2: Operating Procedures (site- or company-specific)

- Document control

- Corrective action
- Calibration

Documentation Level 3: Detailed Information (product- or process-specific)
- Job instructions
- Planning
- Inspection
- Testing information
- Forms, tags, labels

Documentation Level 4: Quality Records (need-specific)
- Forms
- Files
- Books
- Records

The quality manual describes the quality policy and objectives of the organization, identifies the quality responsibilities and relationships of management, and indicates how the many organizations come together to provide products and services that meet customer requirements. The manual will make references to supporting documentation that exists for each of the individual organizations. The steering committee is responsible for developing and disseminating the contents of the quality manual.

Documentation Survey

The documentation survey provides a list of the documents and files that currently exist in the organization. The intent of the ISO preparation is to utilize existing documentation as much as possible.

The survey is a summary of the statements (citations) in ISO 9001 that require documentation or identification of the quality management system. The summary includes all cases where the word "statement" is used in the quality standard as well as implied document requirements as determined from ISO 9004. The ISO quality system standard does not define "document" or set any type of minimum length or format specifications. The contents of the survey form are hereafter described.

ISO Paragraph or Document Required

The paragraphs and survey forms are based on the responsibility matrix. It

Figure 2-3. Paper Trail for ISO 9000 Documentation

identifies records that must be *"identified, collected, indexed, filed and maintained"* as specified in ISO 9001, Paragraph 4.16.

Name of Existing Document
For the purposes of this survey, documents can include policies, procedures, manuals, interoffice memos, organizational charts, flowcharts, historical records, procurement records, customer contact records, complaint records, meeting minutes, standards, product or process descriptions, or process control systems. Hard copies as well as software formats are subject to audit.

Responsible Person or Organization
This identifies responsibility for originating and/or maintaining the document.

Location of the Master
This identifies physical location of the master document, if known. It should be indicated if no master copy is available. This includes both hard copy and software copy.

Data Filed and Location of File
Indicate data and records that must be filed as a result of the requirement, as well as any other files that may assist an auditor in establishing a "documentation trail." Figure 2-3 presents a graphical representation of the essential paper trail for ISO 9000. An auditor will ask to see objective evidence that a requirement is fulfilled. This means that files become very important during the preparation, internal audit, and third-party audit activities. List all applicable files including departmental, personal, databases, and so on.

Document Coordinator

The ISO document coordinator should keep a copy of the documentation survey for his/her files. This will serve as a starting point for preparing for the audit. Also, the paper audit contains several questions that are answered by the documentation survey. Focus should be on the organization, but it should be understood that other coordinators may be gathering information pertaining to the same paragraphs. As far as documentation is concerned, *duplication is not an infraction*. The responsibility matrix will show areas where information overlap may exist for some paragraph assignments. Table 2-2 illustrates contents of an ISO 9000 documentation survey.

Paper Audit

A paper audit is one of the early steps in achieving ISO 9000 certification. Its purpose is to provide a paper trail of functional responsibilities relevant to the ISO quality assurance system. A paper audit complements the usual documentation survey, which focuses on documentation. The paper audit focuses on the organization and relationships between organizations required by the ISO 9001 standard. It allows the site ISO coordinator and the document coordinators in each area to move through the organization assigning appropriate sections to the people responsible for those sections. The questions contained in the paper audit should help explain what the standard requires. The paper audit questionnaire should be used as a tool to identify areas needing attention.

Top site management and ISO 9000 coordinators should use a paper audit in evaluating their areas for readiness for an internal audit prior to a third-party audit for ISO certification. The distribution of the paper audit is limited to all corporate executives, all ISO site coordinators, documentation coordinators, managers, and employees who need to know.

It is the responsibility of all management to assist the site ISO coordinator and documentation coordinators in locating the people responsible for the answers to the questions covered by the paper audit. Both parties must work together to meet the requirements of this standard, thereby ensuring that the site will pass the subsequent audits and meet corporate goals. Some of the questions in a typical paper audit concerning the control of nonconforming product are as follows:

1. Are there written procedures to prevent nonconforming (off-specification) product from inadvertent use or installation? Where are the procedures located? Are the procedures being used?
2. Are affected people (functions) notified of nonconforming product? How is this done?

INTRODUCTION TO ISO 9000   39

TABLE 2-2   ISO 9000 Documentation Survey

Organization: _____
Coordinator: _____

ISO Paragraph or Document Required	Name of Existing Document	Resp. Person or Org.	Location of Master	Data Filed (Record)	Location of File	ISO Readiness (0 to 5)
4.10.4 Procedure for maintaining records which give evidence that product has passed test or inspection						
4.10.4 RECORD Verification data that the product passed inspection or test						
4.12 RECORD Inspection authority for the release of conforming product						
4.14 Documented corrective action procedures						
4.14 RECORD Changes resulting from corrective action						
4.16 Procedures for Identification, collection, indexing, filing, storage, maintenance, and disposition of quality records, including checklists						

3. How is the responsibility and authority for review of nonconforming product defined?
4. Are there written procedures for the review of nonconforming product? Where are they located? Are they being used?
5. How do procedures for review of nonconforming product cover disposition? By rework to meet specified requirements? By scrapping?

Employee Job Descriptions

Employees must be given explicit job descriptions for what they are expected to do. It is a fact that in many organizations, many employees don't have (or are not aware of) written job descriptions for what they do. Ambiguous and nonexistent job statements lead to misinterpretations and misconceptions. Action words must be used in the job descriptions. A concise job description leaves little room for confusion on the part of the worker.

The author witnessed a case situation recently when teenage employees arriving at a fast-food restaurant early in the morning found the doors still locked. They were baffled and didn't know what to do. They had not been given instructions about what to do in such cases, *because it had never happened before*. A manager always got to the restaurant at dawn to open the door before the employees arrived. Well, on this particular day, the manager overslept.

The employees waited by the door for one hour before calling the police, who then called the restaurant owner. Before then, the teenagers simply turned away customers by saying, "We're not open yet." It turned out that the proper thing for the employees to do was to try and reach the manager on duty by telephone. They claimed they did not know which manager was scheduled; they didn't know the managers' home telephone numbers; and, besides, they didn't have coins for pay phones anyway! This may seem like a minor problem. But imagine the customer goodwill that must have been lost that morning, just because employees didn't know what to do in an unusual situation. To avoid this type of problem, employees should be given simple and explicit instructions about potential contingencies. Specific guidelines for job instructions are as follows:

- Keep description simple and concise.
- Specify tools to be used.
- Indicate performance measure for the job.
- Specify what must be done in unusual situations.

- Specify job due date or time schedule.
- Use job flow chart whenever possible.

What is Expected of Employees?

- Be aware of ISO and how it affects your area.
- Follow compliance plan and procedures in your area.
- Correct inaccurate documents in your area.
- Cooperate with auditors and assessors.
- Know who is the quality representative in your area.
- Follow documented work instructions.

SITE IMPLEMENTATION PLAN

A site implementation plan describes the tasks and responsibilities required to achieve ISO 9000 certification. It may be developed for a specific ISO 9000 component (e.g., ISO 9001) and a specific production process. The steering committee should ensure that all managers and documentation coordinators will use this information in evaluating their organization for readiness for an internal audit followed by a third-party audit for ISO certification. Organizational units should be encouraged to dedicate the time and resources needed to accomplish the tasks outlined in the site implementation plan. The steering committee will track the overall progress and report periodically to management. Important components of the implementation plan are

- The responsibility matrix,
- The compliance schedule (timeline), and
- The documentation survey.

ISO POLICY AND PROCEDURES

A quality policy is the overall quality management intentions and direction of an organization as expressed in a formal top-management document. Procedures constitute the vehicle for carrying out a policy. Quality procedures indicate the specific steps required to carry out the policy. A quality policy should be documented in writing, accessible, understandable, and auditable. Quality procedures should be simple to follow, concise and clear, devoid of ambiguities, and verifiable.

42 A TWELVE-MONTH TIMELINE

TASKS	\multicolumn{12}{c	}{TIMELINE (Months)}	Comment
	1	2	3	4	5	6	7	8	9	10	11	12	
Selected employees attend ISO 9000 seminar	▲												
ISO 9001 chosen as standard for site compliance		▲											
Begin coordination with other company sites		▲											
Select lead auditor			▲										
Present ISO 9000 schedule to executive staff				▲									
Form implementation steering team				▲									
Develop timeline					▲								
Steering team and selected executives attend ISO seminar					▲								
Appoint management representative							▲						
Select documentation coordinators							▲						
Distribute ISO information pamphlet								▲					
Select internal auditors and train								▲					
Start ISO awareness training seminars for all employees								▲					
Management representative attend 5-day lead assessor training								▲					
Start internal audit								▲					
Attend documentation seminar								▲					
External consultant conducts ISO gap audit										▲			
Start weekly ISO management review										▲			
Start document center										▲			
First management review of quality system											▲		
External consultant conducts pre-assessment											▲		
Train additional internal auditors in-house												▲	
Second management review of quality system												▲	
Recommend specific product or process for certification													▲
Recommend other product lines for certification (e.g., to ISO 9002)													▲

Figure 2-4. Twelve-Month Timeline for ISO 9000 Process

A TWELVE-MONTH TIMELINE

Activity scheduling is an important part of managing an ISO 9000 process successfully. Figure 2-4 presents a twelve-month timeline for pursuing ISO 9000 certification for a specific product line. The timeline can be easily modified to suit specific company needs. In addition to a general timeline, a gap-analysis worksheet and milestone chart can be used.

TIPS FOR A SUCCESSFUL PROCESS

The principle of KISS (keep it simple, sweetie) should be adopted when implementing the ISO process. Although the standards appear to give rigid guidelines, the implementation is really flexible. Tips for a successful ISO 9000 process are presented:

- Keep the approach simple and flexible.

TABLE 2-3 DO's and DON'Ts of ISO 9000

DO's	DON'Ts
Get management support.	*Don't have outsiders write procedures.*
Allow ISO 9000 to complement existing quality programs.	*Don't leave employees uninformed about the process.*
Survey the work force and management to ensure that questions are being answered and needs are being met.	*Don't expect ISO 9000 registration to solve quality problems. It can only identify areas where solutions are needed.*
Solicit on-the-job volunteers to draft ISO job procedures.	*Don't oversell the idea.*
Review drafts with all staff involved in the procedure to ensure support.	*Don't leave the process unattended.*
Communicate progress to employees regularly to show how the ISO 9000 program has affected organizational performance.	*Don't let up on the enthusiasm.*

- Develop concise work procedures (e.g., use of flowcharts).
- Make necessary adjustment to internal process.
- Use benchmarking approach to see how others have applied ISO 9000.
- Appoint in-house champion of ISO 9000.
- Use outside experts sparingly.
- Involve everyone in the organization.
- Avoid selling ISO 9000 as a rigid set of standards.
- Stay one step beyond minimum requirements.
- Learn, train, practice, and keep up.

The DO's and DON'Ts of ISO 9000 are summarized in Table 2-3.

IMPACT OF TECHNOLOGY CHANGES

The fast pace of technological changes can adversely affect compliance with standards. As technologies change, standards must adapt to reflect new production capabilities and operating procedures. The microcomputer industry is one example where technology changes frequently necessitate new oper-

TABLE 2-4 Technological Changes in Computer Storage Technology

Technology	1979	1993
Areal Density (Mbits/sq in)	1.96	154.23
Storage Capacity (for same size drive)	5.0 Mbytes	2,912 Mbytes
Average Seek Time (msec)	85	9
Maximum Data Throughput (Mbytes/sec)	0.625	20
Spindle Speed (rpm)	3,600	7,200
Average Latency (msec)	8.33	4.17
Smallest Size (inches)	5.25	1.8
Mean Time Between Failures (Hours)	11,000	500,000
Flying Height of Heads (micro inches)	24	3
Cache Size (Kbytes)	0	1,024

ating procedures. For example, Table 2-4 illustrates dramatic changes in computer storage technology between 1979 and 1993. This is a highly competitive industry in which most companies are actively pursuing ISO certification. Work processes, documentations, and quality control procedures must be frequently updated in such an industry to keep up with the technological changes. Many modern industries fall in this fast-paced technology scenario.

3
TQM AND ISO 9000

Total quality management (TQM) is a concept that has emerged recently as a way to achieve a systems approach to quality management. Total quality management refers to a total commitment to quality. This implies an overall integrated approach to all aspects of quality considering all the people, all the hardware, all the software, and all the organizational resources. This requires the total participation of everyone. *High-quality performance should be required from all the people at all times!*

TQM SYSTEM

Figure 3-1 presents a model of components integration in a TQM system. The model recommends the integration of the various subsystems with respect to quality objectives.

In non-TQM organizations, a variety of systems and subsystems exists without an effective interaction. Such disjoint subsystems may include the following:

- Management system
- Manufacturing system
- Design and engineering system
- Management information system
- Financial information system

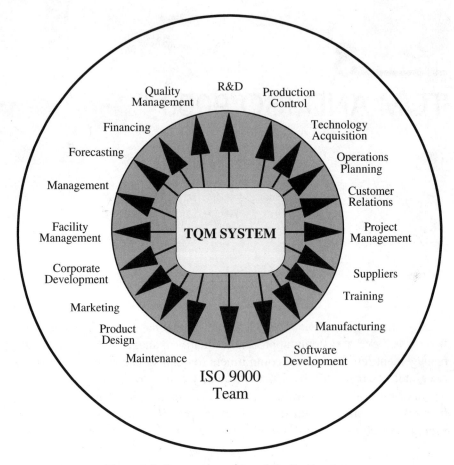

Figure 3-1. Systems Integration of Quality Functions

- Marketing information system
- Inventory information system
- Personnel information system
- Production information system

Some of these subsystems may even have different and conflicting priorities within the same organization. In order to achieve the benefits of TQM, management must ensure that quality objectives are prioritized, integrated, and applied uniformly and consistently throughout the organization with a global systems objective. An organization should pursue a *total commitment to quality*.

Benefits of TQM

Manufacturing and service organizations are under increasing pressure for more effective use of the few resources that are available. Technology and quality management have emerged to play a major role in the success or failure of enterprises. To successfully manage operations, managers will need to view the quality management function on the basis of systems requirements. Specific operations within an organization should be viewed as components of a large system that must be integrated to achieve the overall quality objective of the organization. TQM facilitates an appreciation for integration of advanced technology, a revision of the corporate culture, upgrade of production infrastructure, and better utilization of human resources. It is the people aspects of managing quality that make TQM very essential. TQM is not in effect in an organization if it does not affect all the people and all the products all the time. TQM fosters a new attitude toward quality. TQM offers the following advantages:

- Higher productivity
- Better employee relations
- Cooperative coordination of efforts
- Interaction with subsystem environments
- Uniformity and consistency of quality objectives
- Specification of the interrelationships of subsystems
- Systematic solution of quality problems in an organization
- Coordinated adjustment of functions to solve quality problems
- Dynamic integration of activities into an effective total system
- Increased probability, frequency, and consistency of making good products

JUST IN TIME FOR ISO 9000

Doing things at the right time is the premise of just in time (JIT), a concept that complements the ideas of TQM and facilitates the objectives of ISO 9000. JIT is a materials control strategy that schedules supplies as required for work without buffer stock or excess inventory. JIT calls for getting the right materials in the right quantity at the right time. JIT links a series of work requirements, analyzing demands for the next work to be done. JIT has been called by various names including ZIPS (zero inventory production system), MAN (material as needed), MIPS (minimum inventory production system), stockless production, continuous flow manufacturing, Kanban, Toyota system, and Ohno system (after Taiichi Ohno, a Toyota vice president and mastermind of the system). *What is worth doing is worth doing right, promptly.* The major benefits of JIT can be summarized as follows:

- Fast feedback about process performance
- Reduction in lot size
- More consistent output rates
- Less inventory in system
- Less material waste
- Fewer rework labor hours
- Less indirect cost in the following areas:
  - Interest charge on idle inventory
  - Inventory holding cost
  - Inventory accounting cost
  - Control cost for physical inventory

COMPONENTS OF QUALITY IMPROVEMENT

Management must recognize that quality improvement consists of numerous prerequisite factors. *Little bits of quality make up high quality.* Several factors must be recognized in each specific quality-improvement endeavor. As a minimum, management must play a leadership role and do the following:

- Recognize that *quality takes time.* Rush jobs lead to poor results.
- Recognize that *quality comes from care.*
- Know the customer and get him or her involved very early.
- Simplify the quality-improvement approach. Complicated approaches confuse and discourage participants. Deming's PDCA cycle, for example, is a simple model to implement.
- Recognize that incremental improvements are easier to achieve than one giant improvement.
- Start at the elementary levels of the quality system.
- Use the *divide-and-conquer* approach in organizing quality functions.

CUSTOMER INVOLVEMENT

The absence of defects is no longer a sufficient definition of quality. Thus, the process of improving quality requires that associated operations be

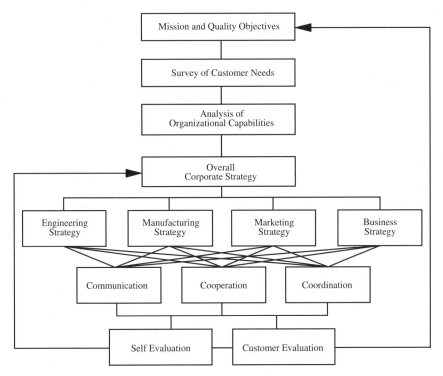

Figure 3-2. Flowchart to Quality Improvement

understood, be improved, and be well-managed. Figure 3-2 presents a flowchart of a quality-improvement process. The business mission and overall quality objectives must be integrated with customer needs with respect to organizational capabilities. There must be a feedback mechanism through which changes in customer needs are conveyed to further drive the mission of the business. Getting closer to the customer and employee empowerment are two of the basic requirements for achieving quality improvement.

Producer-Consumer Partnership

Quality is in the mind of the customer. When pursuing quality improvement, the business and customer definitions of quality must be matched. What the customer wants should be what the producer is willing to provide. A business definition of quality may indicate that *quality is the level of product appeal and functionality needed to make a product acceptable and profitable in the marketplace.* By comparison, the buyer's definition of quality may indicate that *quality is the level of product capability required to satisfy the custom-*

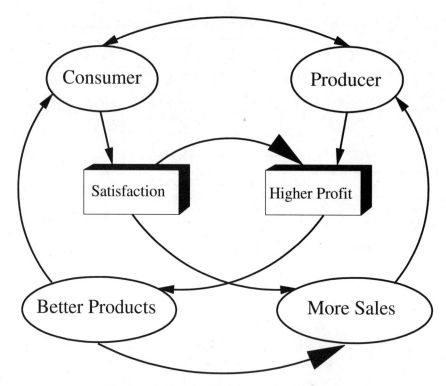

Figure 3-3. Business-Customer Integration Loop

er's use and requirements. The customer may care less whether or not the business is profitable in the provision of the product, whereas the business has to worry about both its profitability requirements and its customers' satisfaction.

The business and customer views should be a two-way affair. If the business is profitable, it will be in a better position to provide better products to meet new customer needs. If the customer is satisfied, he or she will be more willing to embrace the products offered by the business and thereby create further profit potential for the business. One system component must support the other. Quality systems integration requires that the business looks out for the customer while the customer looks out for the survival of the business. For, if there is no business, there will be no product. If there is no product, customers' needs cannot be met. A model of the business-customer integration loop is presented in Figure 3-3.

However, there is a need to correct the conflict between business and customer definitions and views of quality. To enhance the responsiveness to the customer, manufacturers should let customer needs determine product choices.

User Training and Instructions

User training is an important aspect of quality improvement. If a user is given proper training and instructions, the chances of misusing a product will be reduced. To solidify customer involvement in quality improvement, user training is required at the customer end of the transaction. A product that is used and cared for in accordance with the producer's instructions is likely to carry a higher perception of quality. The more time the producer has to spend attending to quality "problems" that result from the misuse of a product, the less time will be available to pursue real improvements in the product.

Consequently, efforts must be made to ensure that products are used correctly. For certain types of products, this cannot be achieved through written instructions only. There must be a combination of instructions and formal training, however short it may be. User training should be the joint responsibility of the user, the customer organization, and the producer. Each has a benefit to be derived from proper product use. In this sense, the customer is defined as the organization that embodies one or more users. By the correct use of a product, the user (an individual) will minimize the chances of frustration with the product. The elimination of frustration will enable him or her to participate more actively in quality-improvement efforts. By the correct use of a product, the customer (an organization) will have more time to devote to real quality-improvement issues. By the correct use of a product, the producer will enjoy more favorable perception of the quality of the product, thus having more time and goodwill to attend to critical quality issues. Because it is impossible to design a product for all kinds of potential abuses, it will actually be cheaper for the producer to participate in user-training programs.

VENDOR CERTIFICATION

A vendor must be committed to the producer; the producer must be committed to the vendor. Just as customers are expected to be involved in quality improvement, so also should vendors be expected to be involved. Customer requirements should be relayed to vendors so that the goods and services they supply to the production process will satisfy what is required to meet customer requirements. Selected vendors may be certified based on their previous records of supplying high-quality products. A comprehensive program of vendor-producer commitment should hold both external vendors and internal production facilities jointly responsible for high-quality products. The importance of vendor involvement is outlined as follows:

- Vendor and producer have a joint understanding of customer requirements.
- Skepticism about a vendor's supply is removed.
- Excessive inspection of a vendor's supply is avoided.
- Cost of inspecting a vendor's supply is reduced.
- Vendors reduce their costs by reducing scrap, rework, and returns.
- Vendor morale is improved by the feeling of participation in the producer's mission.

To facilitate vendor involvement, the producer may assign a liaison to work directly with the vendor in ensuring that the joint quality objectives are achieved. In some cases, the liaison will actually spend time in the vendor's plant. This physical presence helps to solidify the vendor-producer relationship. Also, the technical and managerial capability of the producer can be made available to the vendor for the purpose of source quality improvement. Many large companies have arrangements whereby a team of technical staff is assigned to train and help vendors with their quality-improvement efforts.

Vendor Rating System

A formal system for vendor rating can be useful in encouraging vendor involvement. Vendors that have been certified as supplying high-quality products will enjoy favorable prestige in an organization. The following is an effective vendor rating system that is based on an opinion poll of a team.

Requirements

1. Form a *vendor quality rating team* of individuals who are familiar with the company's operations and the vendor's products.
2. Determine the set of vendors to be included in the rating process.
3. Inform the vendors of the rating process.
4. Each member of the rating team should participate in the rating process.
5. Each member will submit an anonymous evaluation of each vendor based on specified quality criteria.
6. Develop a weighted evaluation of the vendors to arrive at overall relative weights.

TABLE 3-1 Vendor Rating Matrix

	Rating by Member $j=1$	Rating by Member $j=2$	...	...	Rating by Member n	Total pts. for Vendor i	w_i
Rating for Vendor $i=1$							
Rating for Vendor $i=2$							
...							
...							
Rating for Vendor m							
Total pts. from j	100	100	...	...	100	$100n$	

Steps

1. Let T be the total points available to vendors.
2. $T = 100(n)$, where n = number of individuals in the rating team.
3. Rate the performance of each vendor on the basis of specified quality criteria on a scale of 0 to 100.
4. Let x_{ij} be the rating for vendor i by team member j.
5. Let m = number of vendors to be rated.
6. Organize the ratings by team member j as shown below:

    Rating for Vendor 1: x_{1j}
    Rating for Vendor 2: x_{2j}
    .   .
    .   .
    .   .
    Rating for Vendor m: x_{mj}
    Total Rating Points $\underline{100}$

7. Tabulate the team ratings as shown in Table 3-1 and calculate the overall weighted score for each vendor i using the following equation:

$$w_i = \frac{1}{n}\sum_{j=1}^{n} x_{ij}$$

For the case of multiple vendors for the same item, the relative weights, w_i, may be used to determine what fraction of the total supply should be obtained from each vendor. The fraction is calculated as follows:

$$F_i = (w_i)(\text{size of total supply}),$$

where F_i is the fraction of the total supply that should go to vendor i. The size of the supply may be expressed in terms of monetary currency or product units.

EMPLOYEE INVOLVEMENT

Employees are the custodians of quality. Employees make quality improvement possible. Grass-roots commitment to quality must be pursued even at the lowest employee level. It has been widely reported that the quality achievements experienced by Japan were due largely to the individual and collective commitment of employees. With the support of management, employees must play active roles in quality-improvement efforts. An organization should do the following in order to increase employee involvement in quality improvement:

- Establish quality-awareness programs.
- Listen to employees.
- Find out what each employee needs in order to do a better job.
- Encourage everyone (from the custodian to the CEO) to speak the *language of quality*.
- Build a participative model through quality circles.
- Watch for the subtle quality danger signals of employees.
- Be specific about the role of employees in quality improvement.
- Create avenues for employee input and feedback on quality.
- Make employees feel like a part of the quality decision team.
- Recognize employee contributions.
- Follow up on employee ideas quickly and forthrightly.

The *total employee involvement* (TEI) concept has emerged in recent times as one approach to achieving quality improvement. TEI is a management practice that emphasizes the significance of employee contribution to quality improvement in the workplace. The concept provides a practical approach for giving everyone the power to be a manager of a process and be responsible for the continuous improvement of that process. This empowerment facilitates goal-orientated actions by all employees. Total employee involvement gets everyone working together synergistically to pursue a common goal of quality improvement. More job-control power should be given to employees without overstepping management's authority. First, employees should be trained in how to best perform their jobs. Then, they should be given the right tools to perform the jobs. Finally, let them control their own approaches to making use of the training and tools available. Total employee involvement encourages a *job-ownership* feeling that can help maximize the contribution of each employee.

QUALITY OF MANUFACTURED GOODS

The quality of a manufactured product is determined by several elements including the design, the production equipment, the infrastructure that supports the production, and the operators that are involved in the production. Thus, quality is manufactured, not acquired. Product proliferation has been one way by which manufacturers attempt to increase market share and revenue. Unfortunately, the increase in the number of product choices is achieved at the expense of quality. Product complexity, in terms of design, production, and marketing, places a big demand on a company's operations and resources. The company must accommodate both low and high volumes for different product lines, short and long lead times, standard and custom products, stock and order products, and fixed versus variable production setups. In addition, the company must provide support services such as customer service, marketing, and personnel training. Product fragmentation is another source of complexity to both the manufacturer and the consumer. Most consumers are familiar with the introduction of products labeled as *"new, improved."* Although there may be no question about the newness of the products, the improved aspect is often questionable. In order to improve product quality, product complexity and fragmentation must be controlled.

Managing Product Complexity

The rate at which new products are introduced far exceeds the rate at which consumers can adapt to the changes. A product's complexity, inherent quality, and compatibility will ultimately influence customer perception of the overall quality of the product. To evaluate how a product might meet the needs of the customer, the following questions should be addressed:

- What is the purpose of the product?

- What are the characteristics of the users?
- What skill level is required to use the product?
- What customer support services will be provided?
- Will the producer furnish all the product components?
- What, if any, product components will users need to buy separately?
- Where will the product be used?
- What communication facilities are available?

Product Planning

In some cases, a redesign of the product may be necessary to meet customers' quality needs. Product planning is the process of determining which products to offer in a competitive market. From the design and blueprint specification stage to the actual production stage, the needs of the customer should be kept in mind. Product planning requires that new designs be added as new products when justified, while old products are modified or discontinued as appropriate. In general, product planning should address the following three essential elements:

1. **Product characteristics**—This covers product plans, market analysis, specifications, drawings, tests, design review, prototypes, cost procedures, and value engineering.
2. **Product configuration**—This should cover drawing release procedures, design review process, change-order procedures, inspection, production level, technical supervision, quality-control procedures, and product assurance (reliability and maintainability).
3. **Coordination of one product with other products**—This involves integrating the activities involved in the production of one product with other product schedules with respect to quality requirements, resource allocation, and production capabilities.

Product planning should be an iterative process that is continuously reviewed and revised based on the prevailing customer needs. Some of the specific items of focus in product planning are

1. **Materials and supplies:** the generation of reports showing materials and supplies that will be required by a product.
2. **Labor:** the analysis of labor hours required to accomplish good quality work on a product.
3. **Overhead allocation:** the distribution of production overhead based on the current mix of products.

4. **Product tracking:** the tracking of the status of a product with respect to production and quality standards.
5. **Job transfer:** the routing of a job from one product center to another with respect to the preservation of quality.

QUALITY OF SERVICE

Quality of service is an important aspect of the pursuit of ISO 9000. A service is often an intangible product that is provided to a client. Interpersonal skills play a more significant role in the quality of services than they do in the quality of manufactured products. Quality alone is not enough, better customer relations should also be pursued. The service provider must ensure that those directly providing service to the client have the proper tools to adequately discharge their duties. The personal needs and recognition of these individuals are essential to achieving improvement in the quality of services. Quality of service has several dimensions:

- *Reliability:* This deals with the consistency, accuracy, and dependability of service.
- *Promptness:* This involves the timeliness, responsiveness, and willingness to provide service.
- *Competence:* This refers to the adequacy of the skill and knowledge required to deliver service.
- *Access:* This involves the receptiveness of the service provider to customer requests.
- *Courtesy:* This deals with the empathy, respect, consideration, and politeness with which a service is provided.
- *Communication:* This refers to the ability to listen to the customer, keep the customer informed, and accept customer feedback.
- *Comprehension:* This refers to the readiness of the service provider to learn, know, and understand the customer and his or her needs.
- *Credibility:* This refers to the honesty, believability, and reputation of the service provider with respect to the delivery of service.
- *Tools:* This refers to the collection of tangible and physical instruments at the disposal of the service provider.

All of the above dimensions come into play in determining the quality of service. An organization must strive to improve in each dimension in order to achieve overall improvement in quality of service.

MANAGEMENT BY OBJECTIVE

Management by objective (MBO) is the management concept whereby a worker is allowed to take responsibility for the design and performance of a task under controlled conditions. It gives each worker a chance to set his or her own objectives in achieving organizational goals. Workers can monitor their own progress and take corrective actions when needed without management intervention. MBO has the following positive characteristics:

1. It encourages each worker to find better ways of performing his or her job.
2. It avoids over-supervision of self-motivated workers.
3. It helps a worker in becoming better aware of what is expected in his or her job function.
4. It permits timely feedback on worker performance.

However, MBO does have some disadvantages that include possible abuse of the freedom to self-direct and possible disruption of overall coordination of efforts. W. Edwards Deming, the quality expert, advocated the abolishment of MBO. There are various implementations of MBO. There are probably some MBO approaches that adversely affect quality and, thereby, justify the call by some quality gurus for its abolishment. If implemented correctly, MBO can create opportunities for quality improvement.

Management by Exception

Management by exception (MBE) is an after-the-fact management approach to control. Contingency plans are not made, and there is no rigid monitoring. Deviations from expectations are considered to be the exception to the rule. When unacceptable deviations from expectations occur, they are investigated and only then is action taken. The major advantage of MBE is that it lessens the management workload. However, it is a dangerous concept to follow, especially for high-risk products and services. Many of the problems that can develop in quality-improvement efforts are such that after-the-fact corrections are expensive or even impossible. As a result, MBE runs counter to the teachings of continuous quality improvement. The irony of MBE is that managers who operate under crisis management are often recognized as superstars because, they solve problems whenever they develop. On the other hand, managers who prevent problems may be viewed as inactive, because they are rarely seen solving problems.

Management should work toward avoiding quality problems, rather than subjecting the organization to potentially expensive remedies. An organization that operates under the notion "if it's not broken, don't fix it" will never have the opportunity to achieve quality improvement.

ONGOING COMMITMENT

Quality should not be an occasional concern of management; it should be an ongoing commitment. Management must play an active role in implementing the systems approach to TQM. Management must discourage the workforce from cutting corners when it comes to quality. The idea of making products more quickly and cheaply should not supersede the idea of making them better. Although making products more quickly and cheaply is a major determinant of short-term competitive edge, better products are invariably the determinant for long-term survival. If the systems view is implemented effectively, quicker and cheaper products can very well coexist with better and more profitable products.

Management must not only proclaim the need for better quality, but must also commit the necessary resources for it. Investments made for quality today will lead to higher profits in the future. Management must overcome its naiveté about the source of quality problems, which sometimes have their origins in the most unimaginable sources. Management can support TQM with a systems viewpoint by doing the following:

- Raise the level of awareness about the implications of low and higher quality.
- Adopt a supportive quality philosophy.
- Back the philosophy with required resources.
- Make TQM a mandatory requirement throughout the organization.
- Institute periodic quality reporting requirements.
- Establish quality liaisons with clients and suppliers.
- Adopt a flexible perception of systems operations.
- Play a visible role in quality management.
- Require that functional managers document how department-level quality decisions affect other units of the organization.
- Appreciate the limitations of automation in quality management.

Quality improvement requires change. An organization must be prepared for change. Quality improvement must be instituted in every aspect of everything that the organization does. If employees are better prepared for change, then positive changes can be achieved. The "pain but no gain" aspects of quality improvement can be avoided if proper preparations have been made.

The requirements for getting an organization ready for the drastic changes that may be needed by quality improvement include the following:

- Keep everyone informed of the impending changes.
- Get all employees involved.
- Make employees a part of the decision-input mechanism.
- Highlight the benefits of improved quality.
- Train employees in their new job requirements.
- Create an environment for job enrichment.
- Allay fears about potential loss of jobs because of improved quality.
- Promote change as a transition to better things.
- Make changes in small increments.
- Make employees feel that they are the owners of the change.

PREVENTION VERSUS CURE

The efforts devoted to preventing quality problems will always yield more benefits than the costs of detecting and correcting quality problems. Prevention is the remedy for a cure. If prevention works, there will be no need to search for a cure. The detection-and-correction approach to quality management is a crisis-management approach that can actually foment crisis. If quality problems are allowed to develop, even the best detection system may not be able to locate the source of the problem. Even if the problem source is detected, the existing correction system may not be capable of solving the problem. The prudent approach to quality management is the approach that prevents problems from developing. The concept of *laissez-faire* should never be applied to quality management issues. Whether the product or service is "broken" or not, efforts should be made to continuously "fix" and improve it. Prevention has the following advantages:

- It eliminates the cost and time required to detect problems.
- It keeps employees constantly aware of the importance of good quality.
- It precludes the perils of bad quality.
- It promotes "fire safety" rather than "fire fighting."
- It enhances the faith in the production process.

CONTINUOUS PROCESS IMPROVEMENT

Continuous process improvement (CPI) is an approach for obtaining a steady flow of improvement in any process. The term is frequently applied to product quality management functions. However, CPI is also applicable to any goal-oriented process. CPI is a proven method for improving business, management, or technical processes. The method is based on the following key points:

- Early detention of problems
- Identification of opportunities
- Prioritization of opportunities
- Establishment of a conducive decision-making team
- Comprehensive evaluation of procedures
- Review of methods of improvement
- Establishment of long-term improvement goals
- Continuous implementation of improvement actions
- Company-wide adoption of the CPI concept

A steering committee is typically set up to guide the improvement efforts. The typical functions of the steering committee with respect to CPI include the following:

- Determination of organizational goals and objectives
- Communications with personnel
- Team organization
- Allocation or recommendation of resource requirements
- Administration of the CPI procedures
- Education and guidance for company-wide involvement

To illustrate the benefit of CPI, we may consider the two graphical representations in Figures 3-4 and 3-5. The first figure represents the conventional approach to process improvement. The second figure represents the approach of CPI.

In Figure 3-4, the process starts with a certain level of quality. A certain quality level is specified as the target to be achieved by time T. If neglected, the process quality gradually degrades until it falls below the lower control limit at time t_1. At that time, a sudden effort (or innovation) is needed to improve the quality. If neglected once again, the quality goes through another gradual decline until it again falls below the lower control limit at time t_2. Again, a sudden effort is needed to improve the quality. This cycle of *degradation-innovation* may be repeated several times before time T is reached. At time T, a final attempt is made to suddenly boost the process quality to the target level. Unfortunately, it may be too late to achieve the target quality level. There are many disadvantages to this approach:

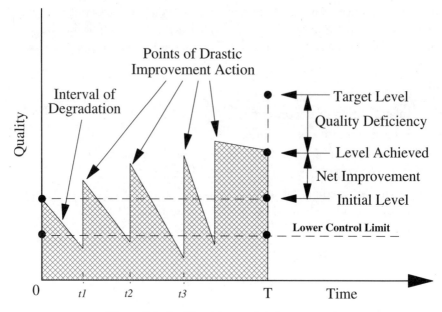

Figure 3-4. Traditional Approach to Improvement

1. High cost of implementation
2. Frequent disruption of the process
3. Too much focus on short-term benefits
4. Need for sudden innovations
5. Opportunity cost during the degradation phase
6. Negative effect on personnel morale
7. Loss of customer trust
8. Need for frequent and strict monitoring
9. Quality control approach rather than quality management approach

In Figure 3-5, the process starts with the same initial quality level and it is continuously improved in a determined pursuit of the target quality level. As opportunities to improve occur, they are immediately implemented. The rate of improvement is not necessarily constant over the planning horizon. Hence, the path of improvement is shown as a curve rather than a straight line. The important aspect of CPI is that each subsequent quality level is at least as good as the one preceding it. As has been mentioned previously, the concept of CPI is applicable to any process. A familiar example is a process whereby students perform better on a test by studying continuously and keeping up with the class rather than trying to study everything the night before the test. The major advantages of CPI include

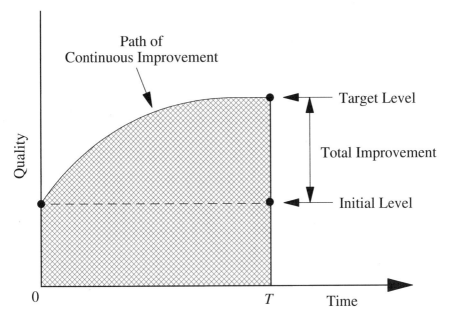

Figure 3-5. Continuous Process Improvement Model

1. Lower cost of achieving quality objectives
2. Better customer satisfaction
3. Dedication to higher-quality products
4. Consistent pace with process technology
5. Conducive environment for personnel involvement
6. Ability to keep ahead of the competition
7. Unambiguous expression of what is expected from the process

A concept related to CPI is the continuous measurable improvement (CMI) approach. CMI is a process through which employees are given the authority to determine how best their jobs can be performed and measured. Because employees are continually in contact with the job, they have the best view of the performance of the process. They have the most reliable criteria for measuring improvements achieved in the process. Under CMI, employees are directly involved in designing the job functions. For example, instead of just bringing in external experts to design a new production line, CMI requires that management get the people (employees) who are going to be using the line involved in the design process. This provides valuable employee insights into the design mechanism and paves the way for the success of the design.

Quality function deployment (QFD) is a planning approach through which an organization evaluates product performance requirements desired by the customer. The effectiveness of QFD comes from translating customer needs and expectations directly into engineering, manufacturing, and service requirements for an organization. Instead of internally setting product characteristics and then hoping that the customer likes what is produced, QFD ensures that what the customer really wants is produced right from the beginning. Some of the advantages of QFD include

- Improved product quality
- Reduction of material waste
- Reduction of cost of rework
- Reduction in product design cycle
- Reduction in product delivery lead time

The implementation of quality function deployment is facilitated by the following guidelines:

1. Conduct credible marketing research
2. Establish a communication liaison with the customer
3. Make the customer aware of the capability and limitation of the production facility
4. Establish cooperation between design, manufacturing, and marketing departments
5. Give feedback to the customer on product design and performance
6. Strive for a trusting relationship with the customer

The QFD process calls for active participation of design, engineering, manufacturing, sales, marketing, and quality assurance groups within the organization. In addition, external contact points must be included.

Pitfall of Ambiguous Customer Surveys

A question not posed is a question not answered. When conducting customer surveys for the purpose of QFD, the questions posed must be explicit and unambiguous. Direct references should be made to the features of the product being surveyed so that useful information may be obtained from customers. For example, in an open-ended survey of customers about the important factors in an airline flight, customers may make reference to such things as quality of food, friendliness of flight attendants, ease of the check-in process, and other obvious factors that they can directly perceive. However, no one may mention anything about the reliability of the airplane's engines, which appear to be the most crucial aspect of flying. QFD surveys must "lead" the respondents in the pertinent direction.

CASE STUDY: TQM IN SOCCER?

The concepts and discipline instilled by TQM and ISO 9000 guidelines can be used to improve any process ranging from recreational activities to professional endeavors. The author has a personal experience in the application of TQM and the ISO 9000 process in coaching a soccer team. The author became, by default, the coach of an adult recreational soccer team (the Crusaders) in the fall of 1992 in the Central Oklahoma Adult Soccer League. Using the TQM and ISO 9000 processes, he took the team from being at the bottom of the league to being the league champion in just three seasons. And this was not due to his coaching acumen, but rather due to the way he motivated the team and made everyone aware of his respective responsibilities on the team. As a coach/player, he applied the techniques to the way he handled things and encouraged the other players to do likewise.

He developed a documentation system that informed the players each week of where the team stood in relation to other teams. Every week, he handed out written notes about what the current objectives were and how they would be addressed. Because of this, he was nicknamed the "Memo Coach." It got to a point where the players got used to being given written assignments, and they would jokingly demand their memo for the week. Copies of graphical representations of the game lineup were given to the players to study prior to each game. Each person had to know his immediate coordination points during a game: Who will provide support for whom, who will cover what area of the field, and so on. He applied TQM to various aspects of the team, including

- Team registration
- Team motivation
- Team communication
- Team cooperation
- Team coordination
- Expected individual commitment
- Player camaraderie
- Field preparation
- Sportsmanship
- Play etiquette

- Game lineup
- Training
- Funding

In the second season under this unconventional coach, the team took third place. The players were excited and motivated, and credited their success to the way the management of the team was handled. So, starting the third season, everyone came out highly charged up to move the team forward to an even better season. Of course, there was the season's inaugural memo waiting for them.

One of the favorite memos handed out to the team was the one that indicated the team's track record (Win-Lose-Draw) dating back 10 years. With this, I was able to motivate the team to move to the higher levels of the league. Traditionally, the team has been viewed as one of the so-so teams in the league. Not a bad team, but not the best either. I convinced the players that, while winning was not everything (particularly in an adult soccer league), it sure would feel better than losing. This was in an over-30 league, tactfully referred to as the Masters League, where most of the players were technical or business professionals.

With the high level of motivation, division of labor, and effective utilization of existing resources (soccer skills, or lack thereof), the team was crowned the league champion for fall 1993. This is not a small feat in a league that contained traditional powerhouses. It is interesting to note that the achievement was made with little or no recruitment of additional "skillful" players, who are in short supply anyway in that league. This shows that with proper management, your existing resources can achieve unprecedented levels of improvement.

4
ISO 9001 QUALITY SYSTEM REQUIREMENTS

ISO 9001 presents a model for quality assurance in design/development, production, installation, and servicing. Because it is the most comprehensive standard, this chapter presents a brief introduction to its essential elements. Further details on it and the other ISO standards can be found in the complete ISO 9000 document. ISO 9001 contains 20 elements and presents a model for quality assurance for firms involved in the design, manufacturing, and installation of products and/or services. It is intended that the standards will normally be adopted in their present form with provision for custom modifications to satisfy specific contractual needs. ISO 9000 provides general guidance for such modifications as well as selection of the appropriate quality assurance model, namely ISO 9001, 9002, or 9003.

ISO 9000 certification is one of the steps toward *total customer satisfaction.* A well-documented, well-managed quality management system that covers all key business and manufacturing functions will keep your company very competitive and result in significant cost savings. It requires discipline, commitment, and consistency that must be followed from workday to workday. Ensuring that basic quality system elements are steadfastly in place in the organization will guarantee that subsequent quality improvement efforts will be on a strong foundation. ISO 9000 should extend throughout the organization to ensure total improvement as shown in Figure 4-1.

The nature and degree of organization, structure, resources, responsibilities, procedures, and processes are essential management decisions affecting quality. It is important that they are documented so that they are understood

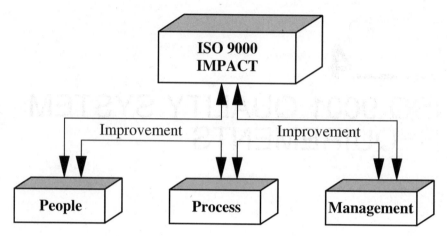

Figure 4-1. Total Improvement through ISO 9000

by the appropriate personnel and the quality system is maintained at a suitable level of detail to provide consistent control of quality.

The quality system must be planned and developed to take into account all supporting functions, such as customer liaison, manufacturing, purchasing, subcontracting, training, and installation. Quality planning must also identify the need for updating quality control techniques, ensure that equipment and personnel are capable of carrying out the plans, and provide adequate quality records.

Cause-and-effect analysis is useful in evaluating the factors that affect quality-improvement efforts. A fishbone diagram is used to develop a relationship between a quality effect and all the possible causes that may create the effect. It can be very effective for organizing and maintaining a quality assurance model to comply with ISO 9000 requirements. Figure 4-2 presents an example of a fishbone diagram. It specifies the relationship between a quality characteristic and a set of factors. The steps for developing a fishbone diagram are presented below:

Step 1: Determine the quality characteristic or the product or process to be studied.

Step 2: Write the characteristic on the right-hand side of a blank sheet of paper. Start with enough room on the paper, because the diagram may expand considerably during the evaluation. Enclose the characteristic in a square. Now, write the primary causes that affect the quality characteristic as big branches (or bones). Enclose the primary causes in squares.

Step 3: Write the secondary causes that affect the big branches as medium-

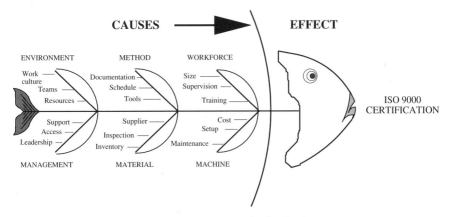

Figure 4-2. Fishbone Diagram for Quality Assurance

sized branches. Write the tertiary (third-level) causes that affect the medium-sized bones as small bones.

Step 4: Assign relative importance ratings to the factors. Mark the particularly important factors that are believed to have a significant effect on the quality characteristic.

Step 5: Append any necessary written documentation to the diagram.

Step 6: Review the overall diagram for completeness. Although it is important to expand the cause-and-effect relationships as much as possible, avoid making the diagram too cluttered. For a fishbone diagram that is to be presented to upper management, limit the contents to a few important details. At the operational level, more details will need to be provided.

ISO 9001 specifies quality system requirements for use when a contract between two parties requires the demonstration of a supplier's capability to design and supply a product. The requirements specified in this standard are aimed primarily at preventing nonconformity at all stages from design to servicing. ISO 9001 is applicable in the following contractual situations:

- When the contract specifically requires design effort and the product requirements are stated principally in performance terms or they need to be established.
- When confidence in product conformance can be attained by adequate demonstration of a certain supplier's capabilities in design, development, production, installation, and servicing.

The section below presents the quality system requirements as presented by ISO 9001. The major requirements cover management responsibility, quality system, contract review, design control, document control, purchasing, purchaser supplied product, product identification and traceability, process control, inspection and testing, inspection, measuring, and test equipment, inspection and test status, control of nonconforming product, corrective action, handling, storage, packaging, and delivery, quality records, internal quality audits, training, servicing, and statistical techniques. Many of these requirements have appropriate subdivisions.

MANAGEMENT RESPONSIBILITY

Quality Policy

The supplier's management shall define and document its policy and objectives for, and commitment to, quality. The supplier shall ensure that this policy is understood, implemented, and maintained at all levels in the organization.

Organization

Responsibility and Authority

The responsibility, authority, and the interrelation of all personnel who manage, perform, and verify work affecting quality shall be defined, particularly for personnel who need the organization freedom and authority to

1. Initiate action to prevent the occurrence of product nonconformity;
2. Identify and record any product quality problems;
3. Initiate, recommend, or provide solutions through designated channels;
4. Verify the implementation of solutions; and
5. Control further processing, delivery, or installation of nonconforming product until the deficiency or unsatisfactory condition has been corrected.

Verification of Resources and Personnel

The supplier shall identify in-house verification requirements, provide adequate resources, and assign trained personnel for verification activities.

Verification activities include inspection, test, and monitoring of the design, production, installation, and servicing of the process and/or product; design reviews and audits of the quality system, processes, and/or product shall be carried out by personnel independent of those having direct responsibility for the work being performed.

Management Representative

The supplier shall appoint a management representative who, irrespective of other responsibilities, shall have defined authority and responsibility for ensuring that the requirements of this standard are implemented and maintained.

Management Review

The quality system adopted to satisfy the requirements of this standard shall be reviewed at appropriate intervals by the supplier's management to ensure its continuing suitability and effectiveness. Records of such reviews shall be maintained. Management reviews normally include assessment of the results of internal audits.

QUALITY SYSTEM DOCUMENTATION

The supplier shall establish and maintain a documented quality system as a means of ensuring that product conforms to specified requirements. This shall include

1. The preparation of documented quality system procedures and instructions in accordance with the requirements of this standard; and
2. The effective implementation of the documented quality system procedures and instructions.

In meeting the specified requirements, timely consideration needs to be given to the following activities:

- The preparation of quality plans and a quality manual in accordance with the specified requirements;
- The identification and acquisition of any controls, processes, inspection equipment, fixtures, total production resources, and skills that may be needed to achieve the required quality;
- The updating, as necessary, of quality control, inspection, and testing techniques, including the development of new instrumentation;
- The identification of any measurement requirement involving capability that exceeds the known state of the art in sufficient time for the needed capability to be developed;
- The clarification of standards of acceptability for all features and requirements, including those that contain a subjective element;
- The compatibility of the design, the production process, installation, inspection and test procedures, and the applicable documentation; and
- The identification and preparation of quality records.

CONTRACT REVIEW

The supplier shall establish and maintain procedures for contract review and for the coordination of these activities. Each contract shall be reviewed by the supplier to ensure that

1. The requirements are adequately defined and documented;
2. Any requirements differing from those in the tender are resolved; and
3. The supplier has the capability to meet contractual requirements.

The contract review activities, interfaces, and communication within the supplier's organization should be coordinated with the purchaser's organization, as appropriate.

Records of all contract reviews shall be maintained.

DESIGN CONTROL

General: The supplier shall establish and maintain procedures to control and verify the design of the product in order to ensure that the specified requirements are met.

Design and Development Planning: The supplier shall draw up plans that identify the responsibility for each design and development activity. The plans shall describe or reference these activities and shall be updated as the design evolves.

The design and verification activities shall be planned and assigned to qualified staff equipped with adequate resources.

Organizational and technical interfaces between different groups shall be identified and the necessary information documented, transmitted, and regularly reviewed.

Design Input: Design input requirements relating to the product shall be identified, documented, and their selection reviewed by the supplier for adequacy. Incomplete, ambiguous, or conflicting requirements shall be resolved with those responsible for drawing up these requirements.

Design Output: Design output shall be documented and expressed in terms of requirements, calculations, and analyses. Design output shall

- Meet the design input requirements;
- Contain or reference acceptance criteria;
- Conform to appropriate regulatory requirements whether or not these have been stated in the input information; and

- Identify those characteristics of the design that are crucial to the safe and proper functioning of the product.

Design Verification: The supplier shall plan, establish, document, and assign to competent personnel functions for verifying the design. Design verification shall establish that design output meets the design input requirements by means of design control measures such as

- Holding and recording design reviews;
- Undertaking qualification tests and demonstration;
- Carrying out alternative calculations; and
- Comparing the new design with a similar proven design, if available.

Design Changes: The supplier shall establish and maintain procedures for the identification, documentation, and appropriate review and approval of all changes and modifications.

DOCUMENT CONTROL

The supplier shall establish and maintain procedures to control all documents and data that relate to the requirements of this standard. These documents shall be reviewed and approved for adequacy by authorized personnel prior to issue. This control shall ensure that

- The pertinent issues of appropriate documents are available at all locations where operations essential to the effective functioning of the quality system are performed; and
- Obsolete documents are promptly removed from all points of issue or use.

Document Changes and Modifications: Changes to documents shall be reviewed and approved by the same functions/organizations that performed the original review and approval unless specifically designated otherwise. The designated organizations shall have access to pertinent background information on which to base their review and approval.

When practicable, the nature of the change shall be identified in the document or appropriate attachments.

A master list or equivalent document control procedure shall be established to identify the current revision of documents in order to preclude the use of nonapplicable documents. Documents shall be reissued after a practical number of changes have been made.

PURCHASING

General: The supplier shall ensure that purchased product conforms to specified requirements.

Assessment of Subcontractors: The supplier shall select subcontractors on the basis of their ability to meet subcontract requirements, including quality requirements. The supplier shall establish and maintain records of acceptable subcontractors.

The selection of subcontractors, and the type and extent of control exercised by the supplier, shall be dependent on the type of product and, when appropriate, on records of subcontractors' previously demonstrated capability and performance. The supplier shall ensure that quality system controls are effective.

Purchasing Data: Purchasing documents shall contain data clearly describing the product ordered, including, when applicable,

- The type, class, style, grade, or other precise identification;
- The title or other positive identification, and applicable issue of specifications, drawings, process requirements, inspection instructions, and other relevant technical data, including requirements for approval or qualification of product, procedures, process equipment and personnel; and
- The title, number, and issue of the quality system standard to be applied to the product.

The supplier shall review and approve purchasing documents for adequacy of specified requirements prior to release.

Verification of Purchased Product: When specified in the contract, the purchaser or the purchaser's representative shall be afforded the right to verify at source or on receipt that purchased product conforms to specified requirements. Verification by the purchaser shall not absolve the supplier of the responsibility to provide acceptable product, nor shall it preclude subsequent rejection.

When the purchaser or the purchaser's representation elects to carry out verification at the subcontractor's plant, such verification shall not be used by the supplier as evidence of effective control of quality by the subcontractor.

PURCHASER-SUPPLIED PRODUCT

The supplier shall establish and maintain procedures for verification, storage, and maintenance of purchaser-supplied product provided for incorporation into the supplies. Any such product that is lost, damaged, or is otherwise unsuitable for use shall be recorded and reported to the purchaser. Verification by the supplier does not absolve the purchaser of the responsibility to provide acceptable product.

PRODUCT IDENTIFICATION AND TRACEABILITY

When appropriate, the supplier shall establish and maintain procedures for identifying the product from applicable drawings, specifications, or other documents during all stages of production, delivery, and installation. The individual product or batch shall have a unique identification whenever traceability is a specified requirement. This identification shall be recorded.

PROCESS CONTROL

General: The supplier shall identify and plan the production and, when applicable, installation processes that directly affect quality and shall ensure that these processes are carried out under controlled conditions. Controlled conditions shall include

- Documented work instructions defining the manner of production and installation, when the absence of such instructions would adversely affect quality, use of suitable production and installation equipment, suitable working environment, compliance with reference standards/ codes, and quality plans;
- Monitoring and control of suitable process and product characteristics during production and installation;
- The approval of processes and equipment, as appropriate; and
- Criteria for workmanship, which shall be stipulated, to the greatest practicable extent, in written standards or by means of representative samples.

Special Processes: These are processes that cannot be fully verified by subsequent inspection and testing of the product and where, for example, processing deficiencies may become apparent only after the product is in use. Accordingly, continuous monitoring and/or compliance with documented procedures is required to ensure that the specified requirements are met. These processes shall be qualified and shall also comply with general requirements. Records shall be maintained for qualified processes, equipment, and personnel, as appropriate.

INSPECTION AND TESTING

Receiving Inspection and Testing: The supplier shall ensure that incoming product is not used or processed until it has been inspected or otherwise verified as conforming to specified requirements. Verification shall be in accordance with the quality plan or documented procedures.

When incoming product is released for urgent production purposes, it shall be positively identified and recorded in order to permit immediate recall and replacement in the event of nonconformance to specified requirements.

In determining the amount and nature of receiving inspection, consideration should be given to the control exercised at source and documented evidence of quality conformance provided.

In-Process Inspection and Testing: The supplier shall perform the following tasks.

- Inspect, test, and identify product as required by the quality plan or documented procedures;
- Establish product conformance to specified requirements by use of process monitoring and control methods;
- Hold product until the required inspection and tests have been completed or necessary reports have been received and verified, except when product is released under positive recall procedures (release under positive recall procedures shall not preclude the activities outlined previously); and
- Identify nonconforming product.

Final Inspection and Testing: The quality plan or documented procedures for final inspection and testing shall require that all specified inspection and tests, including those specified either on receipt of product or in process, have been carried out and that the data meet specified requirements.

The supplier shall carry out all final inspection and testing in accordance with the quality plan or documented procedures to complete the evidence of conformance of the finished product to the specified requirements.

No product shall be dispatched until all the activities specified in the quality plan or documented procedures have been satisfactorily completed and the associated data and documentation are available and authorized.

Inspection and Test Records: The supplier shall establish and maintain records that give evidence that the product has passed inspection and/or test with defined acceptance criteria.

INSPECTION, MEASURING, AND TEST EQUIPMENT

The supplier shall control, calibrate, and maintain inspection, measuring, and test equipment, whether owned by the supplier, on loan, or provided by the purchaser, to demonstrate the conformance of product to the specified requirements. Equipment shall be used in a manner that ensures that measurement uncertainty is known and is consistent with the required measurement capability. The supplier shall

1. Identify the measurements to be made, the accuracy required, and select the appropriate inspection, measuring, and test equipment;
2. Identify, calibrate, and adjust all inspection, measuring and test equipment, and devices that can affect product quality at prescribed intervals, or prior to use, against certified equipment having a known valid relationship to nationally recognized standards. When no such standards exist, the basis used for calibration shall be documented;
3. Establish, document, and maintain calibration procedures, including details of equipment type, identification number, location, frequency of checks, check method, acceptance criteria, and the action to be taken when results are unsatisfactory;
4. Ensure that the inspection, measuring, and test equipment is capable of the accuracy and precision necessary;
5. Identify inspection, measuring, and test equipment with a suitable indicator or approved identification record to show the calibration status;
6. Maintain calibration records for inspection, measuring, and test equipment;
7. Assess and document the validity of previous inspection and test results when inspection, measuring, and test equipment is found to be out of calibration;
8. Ensure that the environmental conditions are suitable for the calibrations, inspections, measurements, and tests being carried out;
9. Ensure that the handling, preservation, and storage of inspection, measuring, and test equipment is such that the accuracy and fitness for use are maintained; and

10. Safeguard inspection, measuring, and test facilities, including both test hardware and test software, from adjustments that would invalidate the calibrations setting.

When test hardware (e.g., jigs, fixtures, templates, patterns) or test software is used as suitable forms of inspection, they shall be checked to prove that they are capable of verifying the acceptability of product prior to release for use during production and installation and shall be rechecked at prescribed intervals. The supplier shall establish the extent and frequency of such checks and shall maintain records as evidence of control. Measurement design data shall be made available, when required by the purchaser or his representative, for verification that they are functionally adequate.

INSPECTION AND TEST STATUS

The inspection and test status of product shall be identified by using markings, authorized stamps, tags, labels, routing cards, inspection records, test software, physical location, or other suitable means, which indicate the conformance or nonconformance of product with regard to inspection and tests performed. The identification of inspection and test status shall be maintained, as necessary, throughout production and installation of the product to ensure that only product that has passed the required inspections and tests is dispatched, used, or installed. Records shall identify the inspection authority responsible for the release of conforming product.

CONTROL OF NONCONFORMING PRODUCT

The supplier shall establish and maintain procedures to ensure that product that does not conform to specified requirements is prevented from inadvertent use or installation. Control shall provide for identification, documentation, evaluation, segregation when practical, disposition of nonconforming product, and for notification to the functions concerned.

Nonconforming Review and Disposition: The responsibility for review and authority for the disposition of nonconforming product shall be defined. Nonconforming product shall be reviewed in accordance with documented procedures. It may be

- Reworked to meet the specified requirements, or
- Accepted with or without repair by concession, or

- Re-graded for alternative applications, or
- Rejected or scrapped.

When required by the contract, the proposed use or repair of product that does not conform to specified requirements shall be reported for concession to the purchaser or the purchaser's representative. The description of nonconformity that has been accepted, and of repairs, shall be recorded to denote the actual condition. Repaired and reworked product shall be re-inspected in accordance with documented procedures.

CORRECTIVE ACTION

The supplier shall establish, document, and maintain procedures for

- Investigating the cause of nonconforming product and the corrective action needed to prevent recurrence;
- Analyzing all processes, work operations, concessions, quality records, service reports, and customer complaints to detect and eliminate potential causes of nonconforming product;
- Initiating preventive actions to deal with problems to a level corresponding to the risks encountered;
- Applying controls to ensure that corrective actions are taken and that they are effective; and
- Implementing and recording changes in procedures resulting from corrective action.

HANDLING, STORAGE, PACKAGING, AND DELIVERY

General: The supplier shall establish, document, and maintain procedures for handling, storage, packaging, and delivery of product.

Handling: The supplier shall provide methods and means of handling that prevent damage or deterioration.

Storage: The supplier shall provide secure storage areas or stock rooms to prevent damage or deterioration of product, pending use or delivery. Appropriate methods for authorizing receipt and the dispatch to and from such areas shall be stipulated. In order to detect deterioration, the condition of product in stock shall be assessed at appropriate intervals.

Packaging: The supplier shall control packing, preservation, and marking processes (including materials used) to the extent necessary to ensure conformance to specified requirements and shall identify, preserve, and segregate all product from the time of receipt until the supplier's responsibility ceases.

Delivery: The supplier shall arrange for the protection of the quality of product after final inspection and test. When contractually specified, this protection shall be extended to include delivery to destination.

QUALITY RECORDS

The supplier shall establish and maintain procedures for identification, collection, indexing, filing, storage, maintenance, and disposition of quality records. Quality records shall be maintained to demonstrate achievement of the required quality and the effective operation of the quality system. Pertinent subcontractor quality records shall be an element of these data.

All quality records shall be legible and identifiable to the product involved. Quality records shall be stored and maintained in such a way that they are readily retrievable in facilities that provide a suitable environment to minimize deterioration or damage and to prevent loss. Retention times of quality records shall be established and recorded. When agreed contractually, quality records shall be made available for evaluation by the purchaser or the purchaser's representative for an agreed-upon period.

INTERNAL QUALITY AUDITS

The supplier shall carry out a comprehensive system of planned and documented internal quality audits to verify whether quality activities comply with planned arrangements and to determine the effectiveness of the quality system. Audits shall be scheduled on the basis of the status and importance of the activity.

The audits and follow-up actions shall be carried out in accordance with documented procedures. The results of the audits shall be documented and brought to the attention of the personnel having responsibility in the area audited. The management personnel responsible for the area shall take timely corrective action on the deficiencies found by the audit.

TRAINING

The supplier shall establish and maintain procedures for identifying the training needs and provide for the training of all personnel performing activities affecting quality. Personnel performing specific assigned tasks shall be qualified on the basis of appropriate education, training, and/or experience, as required. Appropriate records of training shall be maintained.

SERVICING

When servicing is specified in the contract, the supplier shall establish and maintain procedures for performing and verifying that servicing meets the specified requirements.

STATISTICAL TECHNIQUES

When appropriate, the supplier shall establish procedures for identifying adequate statistical techniques required for verifying the acceptability of process capability and product characteristics.

5
ISO 9000 PROJECT MANAGEMENT

The preceding chapters indicate what needs to be done. This chapter presents guidelines for how to do it using project management techniques. Project management represents an excellent basis for integrating various management techniques such as operations research, operations management, forecasting, quality control, and simulation. Traditional approaches to project management use these techniques in a disjointed fashion, thus ignoring the potential interplay among the techniques. *Project management is the process of managing, allocating, and timing resources to achieve a specific goal in an efficient and expedient manner.*

ELEMENTS OF PROJECT MANAGEMENT

Project management continues to grow as an effective means of managing functions in any organization. Project management should be an enterprise-wide endeavor, the application of project management techniques and practices across the full scope of the enterprise. This concept is also referred to as management by projects (MBP). Management by projects is a recent concept that employs project management techniques in various functions within an organization. MBP recommends pursuing endeavors as project-oriented activities. It is an effective way to conduct any business activity. It represents a disciplined approach that defines any work assignment as a project. Under MBP, every undertaking is viewed as a project that must be managed just as any traditional project. The characteristics required of each project so defined are

1. An identified scope and a goal
2. A desired completion time
3. Availability of resources
4. A defined performance measure
5. A measurement scale for review of work

An MBP approach to operations helps in identifying unique entities within functional requirements. This identification helps to determine where functions overlap and how they are interrelated, thus paving the way for better planning, scheduling, and control. Enterprise-wide project management facilitates a unified view of organizational goals and provides a way for project teams to use information generated by other departments to carry out their functions.

The use of project management continues to grow rapidly. The need to develop effective management tools grows with the increasing complexity of new technologies and processes. The life cycle of a new product to be introduced into a competitive market is a good example of a complex process that must be managed with integrative project management approaches. The product will encounter management functions as it goes from one stage to another. Project management will be needed throughout the design and production stages of the product and in developing marketing, transportation, and delivery strategies for the product. When the product finally gets to the customer, project management will be needed to integrate its use with those of other products within the customer's organization.

An integrated project management approach can help diminish the complexity of such needs through good project planning, organizing, scheduling, and control. The project management institute defines the project management body of knowledge as those topics, subject areas, and processes that are used in conjunction with sound project management principles to collectively execute a project. The major functional areas of project management are *scope, quality, time, cost, risk, human resources, contract/procurement,* and *communications.*

Scope management refers to the process of directing and controlling the entire scope of the project with respect to a specific goal. Establishment and clear definition of project goals and objectives form the foundation of scope management. The scope and plans from the baseline against which changes or deviations can be monitored and controlled. A project that is out of scope may be out of luck as far as satisfactory completion is concerned.

Quality management involves ensuring that the performance of a project conforms to specifications with respect to the requirements and expectations of the project stakeholders and participants. The objective of quality management is to minimize deviation from the actual project plans. Quality management must be performed throughout the life cycle of a project, not just by a final inspection of the product.

Time management involves the effective and efficient use of time to facilitate the execution of a project expeditiously. Time is often the most noticeable aspect of a project. Consequently, time management is of utmost importance in project management. The first step of good time management is to develop a project plan that represents the process and techniques needed to execute the project satisfactorily. The effectiveness of time management is reflected in the schedule performance. Hence, scheduling is a major focus in project management.

Cost management is a primary function in project management. Cost is a vital criterion for assessing project performance. Cost management involves having an effective control over project costs through the use of reliable techniques of estimation, forecasting, budgeting, and reporting. Cost estimation requires collecting relevant data needed to estimate elemental costs during the life cycle of a project. Cost planning involves developing an adequate budget for the planned work. Cost control involves a continual process of monitoring, collecting, analyzing, and reporting cost data.

Risk management is the process of identifying, analyzing, and recognizing the various risks and uncertainties that might affect a project. Change can be expected in any project environment. Change involves risk and uncertainty. Risk analysis outlines possible future events and their likelihood of occurrence. With the information from risk analysis, the project team can be better prepared for change with good planning and control actions. By identifying the various project alternatives and their associated risks, the project team can select the most appropriate courses of action.

Human resources management recognizes the fact that people make things happen. Even in highly automated environments, human resources are still a key element in accomplishing goals and objectives. Human resources management involves the function of directing human resources throughout a project's life cycle. This requires behavioral knowledge to achieve project objectives. Employee involvement and empowerment are crucial elements of achieving the quality objectives of a project. The project manager is the key player in human resources management. Good leadership qualities and interpersonal skills are essential for dealing with both internal and external human resources associated with a project. The legal and safety aspects of employee welfare are important factors in human resources management.

Contract/procurement management involves the process of acquiring the necessary equipment, tools, goods, services, and resources needed to successfully accomplish project goals. The buy, lease, or make options available to the project must be evaluated with respect to time, cost, and technical performance requirements. Contractual agreements constitute the legal documents that define the work obligations of each participant in a project. Procurement refers to the actual process of obtaining the needed services and resources.

Communications management refers to the functional interface between individuals and groups within the project environment. This involves proper organization, routing, and control of information needed to facilitate work. Good communication is in effect when there is a common understanding of information between the communicator and the target. Communications management facilitates unity of purpose in the project environment. The success of a project is directly related to the effectiveness of project communication. From the author's experience, most project problems can be traced to a lack of proper communication.

STEPS OF PROJECT MANAGEMENT

Project management consists of four major phases that encompass several steps:

1. Planning
2. Organizing
3. Scheduling
4. Control

These four phases must be kept in mind in all project management endeavors. The phases can be expanded to cover specific steps as depicted in Figure 5-1. The steps start from problem definition and end with project termination. Problem definition, mission statement, and planning can be categorized under the general phase of Planning. Organizing and resource allocation can be classified under Organizing. Scheduling and tracking can go under the Scheduling phase. Reporting, control, and termination belong to the Control phase.

Overview

Problem Identification
This is the stage where a need for a proposed project is identified, defined, and justified. A project may be concerned with the development of new products, implementation of new processes, or improvement of existing facilities.

Project Definition
Project definition is the phase at which the purpose of the project is clarified.

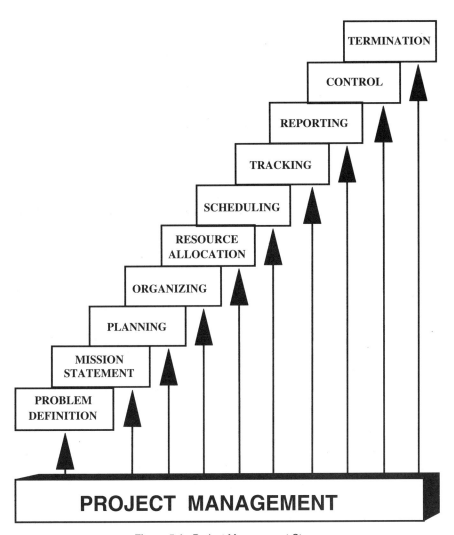

Figure 5-1. Project Management Steps

A *mission statement* is the major output of this stage. For example, a prevailing low level of productivity may indicate a need for a new manufacturing technology. In general, the definition should specify how project management may be used to avoid missed deadlines, poor scheduling, inadequate resource allocation, lack of coordination, poor quality, and conflicting priorities.

Project Planning

A plan represents the outline of the series actions needed to accomplish a goal. Project planning determines how to initiate a project and execute its objectives. It may be a simple statement of a project goal or it may be a detailed account of procedures to be followed during the project. Planning can be summarized as

- Objectives
- Project definition
- Team organization
- Performance criteria (time, cost, quality)

Project Organizing

Project organization specifies how to integrate the functions of the personnel involved in a project. Organizing is usually done concurrently with project planning. Directing is an important aspect of project organization that involves guiding and supervising the project personnel. It is a crucial aspect of the management function. Directing requires skillful managers who can interact with subordinates effectively through good communication and motivation techniques. A good project manager would facilitate project success by directing his/her staff, through proper task assignments, toward the project goal. Workers perform better when there are clearly defined expectations. They need to know how their job functions contribute to the overall goals of the project. Workers should be given some flexibility for self-direction in performing their functions. Individual worker needs and limitations should be recognized by the manager when directing project functions. Directing a project requires skills dealing with motivating, supervising, and delegating.

Resource Allocation

Project goals and objectives are accomplished by allocating resources to functional requirements. Resources can consist of money, people, equipment, tools, facilities, information, skills, and so on. These are usually in short supply. The people needed for a particular task may be committed to other ongoing projects. A crucial piece of equipment may be under the control of another team.

Proper Scheduling

Timeliness is the essence of project management. Scheduling is often the major focus in project management. The main purpose of scheduling is to allocate resources so that the overall project objectives are achieved within a reasonable time span. Project objectives are generally conflicting in nature. For example, minimization of the project completion time and minimization of the project cost are conflicting objectives. That is, one objective is improved at the expense of the other objective. Therefore, project scheduling is a multiple-objective decision-making problem.

In general, scheduling involves the assignment of time periods to specific tasks within the work schedule. Resource availability, time limitations, urgency level, required performance level, precedence requirements, work priorities, technical constraints, and other factors complicate the scheduling process. Thus, the assignment of a time slot to a task does not necessarily ensure that the task will be performed satisfactorily in accordance with the schedule. Consequently, careful control must be developed and maintained throughout the project scheduling process. Project scheduling involves resource availability analysis (human, material, money) and scheduling techniques (CPM, PERT, Gantt charts).

Project Tracking and Reporting

This phase involves checking whether or not project results conform to project plans and performance specifications. Tracking and reporting are prerequisites for project control. A properly organized report of the project status will help identify any deficiencies in the progress of the project and help pinpoint corrective actions.

Project Control

Project control requires that appropriate actions be taken to correct unacceptable deviations from expected performance. Control is actuated through measurement, evaluation, and corrective action. Measurement is the process of evaluating the relationship between planned performance and actual performance with respect to project objectives. The variables to be measured, the measurement scales, and the measuring approaches should be clearly specified during the planning stage. Corrective actions may involve rescheduling, re-allocation of resources, or expedition of task performance. Control involves tracking and reporting, measurement and evaluation, and corrective action (plan revision, rescheduling, updating).

Project Termination

This is the last stage of a project. The phaseout of a project is as important as its initiation. A project should not be allowed to drag on after the expected completion time. A terminal activity should be defined for a project during the planning phase. An example of a terminal activity may be the submission of a final report, switching on new equipment, or the signing of a release order. The conclusion of such an activity should be viewed as the completion of the project. Arrangements may be made for follow-up activities that may improve or extend the outcome of the project. These follow-up or spin-off projects should be managed as new projects but with proper input-output relationships within the sequence of projects.

SELECTING THE PROJECT MANAGER

The role of a manager is to use available resources (manpower and tools) to accomplish goals and objectives. A project manager has the primary responsibility of ensuring that a project is implemented according to the project plan. The project manager has a wide span of interaction within and outside the project environment. He or she must be versatile, assertive, and effective in handling problems that develop during the execution phase of the project. Selecting a project manager requires careful consideration; it is one of the most crucial project functions. The project manager should be someone who can get the job done promptly and satisfactorily. He or she should possess both technical and administrative credibility, and must be perceived as having the knowledge to direct the project. He or she must be up-to-date on the technologies pertinent to the project requirements, and must be conversant in the industry's terminologies. The project manager must also be a good recordkeeper. Because the project manager is the vital link between the project and upper management, he or she must be able to convey information at various levels of detail. The project manager should have good leadership qualities. Because leadership is an after-the-fact attribute, caution should be exercised in extrapolating prior observations to future performance when evaluating candidates for the post of project manager.

The selection process should be as formal as a regular recruiting process. A pool of candidates may be developed through nominations, applications, eligibility records, shortlist, or administrative appointment. The candidates should be aware of the nature of the project and what they would be expected to do. Formal interviews may be required in some cases, particularly those involving large projects. In a few cases, the selection may have to be made by default if there are no other suitably qualified candidates. Default appointment of a project manager implies that no formal evaluation process was carried out. Political considerations and quota requirements often lead to default selection of project managers. As soon as a selection is made, an announcement should be made to inform the project team of the decision. The desirable attributes of a project manager are

- Inquisitiveness
- Good labor relations
- Good motivational skills
- Availability and accessibility
- Versatility with company operations
- Good rapport with senior executives
- Good analytical and technical background

- Technical and administrative credibility
- Perseverance toward project goals
- Excellent communication skills
- Receptive ear for suggestions
- Good leadership qualities
- Good diplomatic skills
- Congenial personality

SELLING THE PROJECT PLAN

The project plan must be sold throughout the organization. Different levels of detail will be needed when presenting the project to various groups in the organization. The higher the level of management, the lower the level of detail. Top management will be more interested in the global aspects of the project. For example, when presenting the project to management, it is necessary to specify how the overall organization will be affected by the project. When presenting the project to the supervisory-level staff, the most important aspect of the project may be the operational level of detail. At the worker or operator level, the individual will be more concerned about how he or she fits into the project. The project manager or analyst must be able to accommodate these various levels of detail when presenting the plan to both participants in and customers of the project. Regardless of the group being addressed, the project presentation should cover the following elements at the appropriate levels of detail:

- Executive summary
- Introduction
- Project description (goals and objectives, expected outcome)
- Performance measures
- Conclusion

The use of charts, figures, and tables is necessary for better communication with the management. A presentation to middle-level managers may follow the more detailed outline that follows:

- Objectives
- Methodologies
- What has been done
- What is currently being done

- What remains to be done
- Problems encountered to date
- Results obtained to date
- Future work plan
- Conclusions and recommendations

STAFFING THE PROJECT

Once the project manager has been selected and formally installed, one of his or her first tasks is the selection of personnel for the project. In some cases, the project manager simply inherits a project team that has been formed before he or she was selected as the project manager. In that case, the project manager's initial responsibility will be to ensure that a good project team has been formed. The project team should be chosen on the basis of skills relevant to the project requirements and team congeniality. The personnel required may be obtained either from within the organization or from outside sources. If outside sources are used, a clear statement should be made about the duration of the project assignment. If opportunities for permanent absorption into the organization exist, the project manager may use that fact as an incentive both in recruiting for the project and running the project. An incentive for internal personnel may be the opportunity for advancement within the organization.

Job descriptions should be prepared in unambiguous terms. Formal employment announcements may be issued or direct contacts through functional departments may be used. The objective is to avoid having a pool of applicants that is either too large or too small. If job descriptions are overly broad, many unqualified people will apply. If the descriptions are too restrictive, very few of those qualified will apply. Some skill tolerance or allowance should be established. As it is nearly impossible to obtain the perfect candidate for each position, some preparation should be made for in-house specialized-skill development to satisfy project objectives. Typical job classifications in a project environment include

- Project administrator
- Project director
- Project coordinator
- Program manager
- Project manager
- Project engineer
- Project assistant

- Project specialist
- Task manager
- Project auditor

Staff selection criteria should be based on project requirements and the availability of a staff pool. Factors to consider in staff selection include

- Recommendation letters and references
- Salary requirements
- Geographical preference
- Education and experience
- Past project performance
- Time frame of availability
- Frequency of previous job changes
- Versatility for project requirements
- Completeness and directness of responses
- Special project requirements (quotas, politics, etc.)
- Overqualification (overqualified workers tend to be unhappy at lower job levels)
- Organizational skills

An initial screening of the applicants on the basis of the factors above may help reduce the applicant pool to a manageable level. If company policy permits, direct contact over the telephone or in person may then be used to further prune the pool of applicants. A direct conversation usually brings out more accurate information about applicants. In many cases, people fill out applications by writing what they feel the employer wants to read, rather than what they want to say. Direct contact can help to find out if the applicant is really interested in the job, whether he or she will be available when needed, or whether he or she possesses vital communications skills.

Confidentiality of applications should be maintained, particularly for applicants who do not want a disclosure to their current employers. References should be checked out and the information obtained should be used with utmost discretion. Interviews should then be arranged for the leading candidates. Final selection should be based on the overall merits of the applicants, rather than on mere personality appeal. Both the successful and the unsuccessful candidates should be informed of the outcome as soon as administrative policies permit.

In many technical fields, manpower shortage is a serious problem. The problem of recruiting in such circumstances becomes that of expanding the pool of applicants rather than pruning it. It is a big battle among employers to entice highly qualified technical personnel from one another. Some recruiters have even been known to resort to unethical means in the attempt to lure prospective employees. Project staffing involving skilled manpower can be enhanced by the following:

- Employee exchange programs
- Transfers from other projects
- In-house training for new employees
- Use of temporary project consultants
- Diversification of in-house job skills
- Cooperative arrangements between employers
- Continuing education for present employees

Committees may be set up to guide the project effort from the recruitment stage to the final implementation stage. The primary role of a project committee should be to provide supporting consultations to the project manager. Such a committee might use the steering committee model that is formed by including representatives from different functional areas. The steering committee should serve as an advisory board for the project. A committee may be set up under one of the following two structures:

1. *Ad hoc committee*. This is set up for more immediate and specific purpose (e.g., project feasibility study).
2. *Standing committee*. This is set up on a more permanent basis to oversee ongoing project activities.

ISO 9000 PROJECT DECISION PROCESS

To register or not to register; that's a big question faced by present business organizations. The usual decision steps for project management are applicable to the ISO 9000 decision process. The steps facilitate a proper consideration of the essential elements of decisions in a project environment. These essential elements include problem statement, information, performance measure, decision model, and an implementation of the decision.

Project Decision Steps

Problem Statement
A problem involves choosing between competing, and probably conflicting,

alternatives. The components of problem solving in project management include

- Describing the problem (goals, performance measures)
- Defining a model to represent the problem
- Solving the model
- Testing the solution
- Implementing and maintaining the solution

Problem definition is very crucial. In many cases, *symptoms* of a problem are recognized more readily than its *cause* and *location*. Even after the problem is accurately identified and defined, a benefit/cost analysis may be needed to determine if the cost of solving the problem is justified.

Data and Information Requirements

Information is the driving force for the project decision process; it clarifies the relative states of past, present, and future events. The collection, storage, retrieval, organization, and processing of raw data are important components for generating information. Without data, there can be no information. Without good information, there cannot be a valid decision. The essential requirements for generating information are as follows:

- Ensure than an effective data collection procedure is followed.
- Determine the type and the appropriate amount of data to collect.
- Evaluate the data collected with respect to information potential.
- Evaluate the cost of collecting the required data.

For example, suppose a manager is presented with a recorded fact that says, *"sales for the last quarter are 10,000 units."* This constitutes ordinary data. There are many ways of using this data to make a decision depending on the manager's value system. An analyst, however, can ensure the proper use of the data by transforming it into information, such as *"sales of 10,000 units for last quarter are within x percent of the targeted value."* This type of information is more useful to the manager for decision making.

Performance Measure

A performance measure for the competing alternatives should be specified. The decision maker assigns a perceived worth or value to the available alternatives. Setting measures of performance is crucial to the process of defining and selecting alternatives. Some performance measures commonly used in project management are project cost, completion time, resource usage, and stability in the workforce.

Decision Model

A decision model provides the basis for the analysis and synthesis of information, and is the mechanism by which competing alternatives are compared. To be effective, a decision model must be based on a systematic and logical framework for guiding project decisions. A decision model can be a verbal, graphical, or mathematical representation of the ideas in the decision-making process. A project decision model should have the following characteristics:

- Simplified representation of the actual situation
- Explanation and prediction of the actual situation
- Validity and appropriateness
- Applicability to similar problems

The formulation of a decision model involves three essential components:

1. *Abstraction:* determining the relevant factors
2. *Construction:* combining the factors into a logical model
3. *Validation:* assuring that the model adequately represents the problem

The basic types of decision models for project management are as follows:

Descriptive models: These are directed at describing a decision scenario and identifying the associated problem. For example, a project analyst might use a CPM network model to identify bottleneck tasks in a project.

Prescriptive models: These furnish procedural guidelines for implementing actions. The triple C approach, for example, is a model that prescribes the procedures for achieving communication, cooperation, and coordination in a project environment.

Predictive models: These models are used to predict future events in a problem environment. They are typically based on historical data about the problem situation. For example, a regression model based on past data may be used to predict future productivity gains associated with expected levels of resource allocation. Simulation models can be used when there are uncertainties in the task durations or resource requirements.

"Satisficing" models: These are models that provide trade-off strategies for achieving a satisfactory solution to a problem within given constraints. Goal programming and other multi-criteria techniques provide good satisficing solutions. These models are helpful in cases where time limitation, resource shortage, and performance requirements constrain the implementation of a project.

Optimization models: These models are designed to find the best available solution to a problem subject to a certain set of constraints. For example, a linear programming model can be used to determine the optimal product mix in a production environment.

In many situations, two or more of the above models may be involved in the solution of a problem. For example, a descriptive model might provide insights into the nature of the problem; an optimization model might provide the optimal set of actions to take in solving the problem; a satisficing model might temper the optimal solution with reality; a prescriptive model might suggest the procedures for implementing the selected solution; and a predictive model might predict the expected outcome of implementing the solution.

Making the Decision

Using the available data, information, and the decision model, the decision maker will determine the real-world actions that are needed to solve the stated problem. A sensitivity analysis may be useful for determining what changes in parameter values might cause a change in the decision.

Implementing the Decision

A decision represents the selection of an alternative that satisfies the objective stated in the problem statement. A good decision is useless until it is implemented. An important aspect of a decision is specifying how it is to be implemented. Selling the decision and the project to management requires a well-organized, persuasive presentation. The way a decision is presented can directly influence whether or not it is adopted. The presentation should include at least the following: an executive summary, technical aspects of the decision, managerial aspects of the decision, resources required to implement the decision, the cost of the decision, the time frame for implementing the decision, and the risks associated with the decision.

CONDUCTING ISO 9000 MEETINGS

Meetings are one avenue for information flow for project decision making. Effective management of meetings is an important skill for any managerial staff. Employees often feel that meetings waste time and obstruct productivity. This is because most meetings are poorly organized, improperly managed, called at the wrong time, or even unnecessary. In some organizations, meetings are conducted as a matter of routine requirement rather than necessity. Meetings are essential for communication and decision making. Unfortunately, many meetings accomplish nothing and waste everyone's time. A meeting of 30 people wasting only 30 minutes in effect wastes 15 full hours of employee time. That much time, in a corporate setting, may amount to thousands of dollars in lost time. It does not make sense to use a one-hour meeting to discuss a task that will take only five minutes to perform. That is like hiring someone at a $50,000 annual salary to manage an annual budget of $20,000. Stewart (1993), in a humorous column about management meetings, writes:

"Management meetings are rapidly becoming this country's biggest growth industry. As nearly as I can determine, the working day of a typical middle manager consists of seven hours of *meetings,* plus lunch. Half a dozen years ago at my newspaper, we hired a new middle management editor with an impressive reputation. Unfortunately, I haven't met her yet. On her first day at work, she went into a *meeting* and has never come out." Stewart concludes his satire with, "I'm expected to attend the next meeting. I'm not sure when it's scheduled exactly. I think they're having a meeting this afternoon about that."

In the past, when an employee had a request, he or she went to the boss, who would say either "yes" or "no." The whole process might take less than one minute out of the employee's day. Today, several hierarchies of meetings may need to be held to review the request. Thus, we may have a departmental meeting, a middle management staff meeting, an upper management meeting, an executive meeting, a steering committee meeting, an ad hoc committee meeting, and a meeting with outside consultants all for the purpose of reviewing that simple request. The following is a list of points about project meetings:

- Most of the information passed out at meetings can be more effectively disseminated through ordinary memos. The proliferation of desktop computers and electronic mail should be fully exploited to replace most meetings.
- The point of diminishing returns for any meeting is equal to the number of people that are actually needed for the meeting. The more people at a meeting, the lower the meeting's productivity. The extra attendees only serve to generate unconstructive and conflicting ideas that impede the meeting.
- Not being invited to a meeting could be viewed as an indication of the high value placed on an individual's time within the organization.
- Regularly scheduled meetings with specific time slots often become a forum for social assemblies.
- The optimal adjournment time of a meeting is equal to the scheduled starting time plus five times the number of agenda items minus the start-up time. Mathematically, this is expressed as
  $L = (T + 5N) - S$

where

L = optimal length
T = scheduled time
N = number of agenda items
S = meeting start-up time (i.e., the time taken to actually call the meeting to order)

Since it is difficult to do away with meetings (the necessary and the unnecessary), we must attempt to maximize their output. Some guidelines for running meetings more effectively are hereby presented:

1. Do pre-meeting homework.
    - List topics to be discussed (agenda).
    - Establish the desired outcome for each topic.
    - Determine how the outcome will be verified.
    - Determine who really needs to attend the meeting.
    - Evaluate the suitability of meeting time and venue.
    - Categorize meeting topics (e.g., announcements, important, urgent).
    - Assign time duration to each topic.
    - Verify that the meeting is really needed.
    - Consider alternatives to the meeting (e.g., memo, telephone, electronic mail).
2. Circulate written agenda prior to the meeting.
3. Start meeting on time.
4. Review agenda at the beginning.
5. Get everyone involved; if necessary employ direct questions and eye contact.
6. Keep to the agenda; do not add new items unless absolutely essential.
7. Be a facilitator for meeting discussions.
8. Quickly terminate conflicts that develop from routine discussions.
9. Redirect irrelevant discussions back to the topic of the meeting.
10. Retain leadership and control of the meeting.
11. Recap the accomplishments of each topic before going on to the next. Let those who have made commitments (e.g., promise to look into certain issues) know what is expected of them.
12. End meeting on time.
13. Prepare and distribute minutes. Emphasize the outcome and success of the meeting.

Huyler and Crosby (1993) present an analysis of the economic impact of poorly managed meetings. They provide guidelines for project managers to improve meetings, and they suggest evaluating meetings by asking the following post-meeting questions:

- What did we do well in this meeting?
- What can we improve next time?

Despite the shortcomings of poorly managed meetings, meetings offer a suitable venue for group decision making.

GROUP DECISION MAKING

Many decision situations are complex and poorly understood. No one person has all the information to make all decisions accurately. As a result, crucial decisions are made by a group of people. Some organizations use outside consultants with appropriate expertise to make recommendations for important decisions. Other organizations set up their own internal consulting groups without having to go outside the organization. Decisions can be made through linear responsibility, in which case one person makes the final decision based on inputs from other people. Decisions can also be made through shared responsibility, in which case a group of people share the responsibility for making joint decisions. The major advantages of group decision making are as follows:

1. *Ability to share experience, knowledge, and resources:* Many heads are better than one. A group will possess greater collective ability to solve a given decision problem.
2. *Increased credibility:* Decisions made by a group of people often carry more weight in an organization.
3. *Improved morale:* Personnel morale can be positively influenced because many people have the opportunity to participate in the decision-making process.
4. *Better reasoning:* The opportunity to observe other people's views can lead to an improvement in an individual's reasoning process.

Some disadvantages of group decision making are as follows:

1. It is harder to arrive at a decision. Individuals may have conflicting objectives. For example, while an accounting department representative may want to minimize costs, a labor union representative may

push for more use of the workforce and protest the use of automation to replace workers.
2. Members who oppose the decision will be reluctant in carrying out the decision.
3. If the tone of the discussions is not controlled, future relations between the members of the group will be hampered.
4. There is a loss of productive employee time.

Various Methods

Brainstorming

Brainstorming is a way of generating many new ideas. In brainstorming, the decision group comes together to discuss alternate ways of solving a problem. The members of the brainstorming group may be from different departments, may have different backgrounds and training, and may not even know one another. The diversity of the participants helps to create a stimulating environment for generating different ideas from disparate viewpoints. The technique encourages free, outward expression of new ideas, no matter how far-fetched the ideas might appear. No criticism of any new idea is permitted during the brainstorming session. A major concern in brainstorming is that extroverts may take control of the discussions. For this reason, an experienced and respected individual should manage the brainstorming discussions. The group leader establishes the procedure for proposing ideas, keeps the discussions in line with the group's mission, discourages disruptive statements, and encourages the participation of all members.

After the group runs out of ideas, open discussions are held to weed out the unsuitable ones. It is expected that even the rejected ideas may stimulate the generation of other ideas, which may eventually lead to other favored ideas. Guidelines for improving brainstorming sessions are as follows:

- Focus on a specific decision problem.
- Keep ideas relevant to the intended decision.
- Be receptive to *all* new ideas.
- Evaluate the ideas on a relative basis after exhausting new ideas.
- Maintain an atmosphere conducive to cooperative discussions.
- Maintain a record of the ideas generated.

Delphi Method

The traditional approach to group decision making is to obtain the opinion of experienced participants through open discussions. An attempt is made to reach a consensus among the participants. However, open group discussions are often biased because of the influence or subtle intimidation from dominant individuals. Even when the threat of a dominant individual is not present, opinions may still be swayed by group pressure. This is called the "bandwagon effect" of group decision making.

The Delphi method attempts to overcome these difficulties by requiring individuals to present their opinions anonymously through an intermediary. The method differs from the other interactive group methods because it eliminates face-to-face confrontations. It was originally developed for forecasting applications. But it has been modified in various ways for application to different types of decision making. The method can be quite useful for project management decisions. It is particularly effective when decisions must be based on a broad set of factors. The Delphi method is normally implemented as follows:

1. *Problem definition:* A decision problem that is considered significant is identified and clearly described.
2. *Group selection:* An appropriate group of experts or experienced individuals is formed to address the particular decision problem. Both internal and external experts may be involved in the Delphi process. A leading individual is appointed to serve as the administrator of the decision process. The group may operate through the mail or gather together in a room. In either case, all opinions are expressed anonymously on paper. If the group meets in the same room, care should be taken to provide enough room so that each member does not have the feeling that someone may accidentally or deliberately observe his or her responses.
3. *Initial opinion poll:* The technique is initiated by describing the problem to be addressed in unambiguous terms. The group members are requested to submit a list of major areas of concern in their specialty areas as they relate to the decision problem.
4. *Questionnaire design and distribution:* Questionnaires are prepared to address the areas of concern related to the decision problem. The written responses to the questionnaires are collected and organized by the administrator, who aggregates the responses in a statistical format. For example, the average, mode, and median of the responses may be computed. This analysis is distributed to the decision group. Each member can then see how his or her responses compare with the anonymous views of the other members.
5. *Iterative balloting:* Additional questionnaires based on the previous responses are passed to the members. The members submit their

responses again. They may choose to alter or not alter their previous responses.

6. *Silent discussions and consensus:* The iterative balloting may involve anonymous written discussions of why some responses are correct or incorrect. The process is continued until a consensus is reached. A consensus may be declared after five or six iterations of the balloting or when a specified percentage (e.g., 80 percent) of the group agrees on the questionnaires. If a consensus cannot be declared on a particular point, it may be displayed to the whole group with a note that it does not represent a consensus.

In addition to its use in technological forecasting, the Delphi method has been widely used in other general decision making. Its major characteristics of anonymity of responses, statistical summary of responses, and controlled procedure make it a reliable mechanism for obtaining numeric data from subjective opinion. The major limitations of the Delphi method are:

- Its effectiveness may be limited in cultures where strict hierarchy, seniority, and age influence decision-making processes.
- Some experts may not readily accept the contributions of non-experts to the group decision-making process.
- Because opinions are expressed anonymously, some members may take the liberty to make ludicrous statements. However, if the group composition is carefully reviewed, this problem may be avoided.

Nominal Group Technique

Nominal group technique is a silent version of brainstorming. It is a method of reaching consensus. Rather than asking people to state their ideas aloud, the team leader asks each member to jot down a minimum number of ideas, for example, five or six. A single list of ideas is then composed on a chalkboard for the whole group to see. The group then discusses the ideas and weeds out some iteratively until a final decision is made. The nominal group technique is easier to control. Unlike brainstorming where members may get into shouting matches, it permits members to silently present their views. In addition, it allows introverted members to contribute to the decision without the pressure of having to speak out too often.

In all of the group decision-making techniques, an important aspect that can enhance and expedite the decision-making process is to require that members review all pertinent data before coming to the group meeting. This will ensure that the decision process is not impeded by trivial preliminary discussions. Some disadvantages of group decision making are:

1. Peer pressure in a group situation may influence a member's opinion or discussions.
2. In a large group, some members may not get to participate effectively in the discussions.
3. A member's relative reputation in the group may influence how well his or her opinion is rated.
4. A member with a dominant personality may overwhelm the other members in the discussions.
5. The limited time available to the group may create a time pressure that forces some members to present their opinions without fully evaluating the ramifications of the available data.
6. It is often difficult to get all members of a decision group together at the same time.

Despite the noted disadvantages, group decision making definitely has many advantages that may nullify the shortcomings. The advantages as presented earlier will have varying levels of effect from one organization to another. The triple C model may be used to improve the success of decision teams. Teamwork can be enhanced in group decision making by following these guidelines:

1. Get a willing group of people together.
2. Set an achievable goal for the group.
3. Determine the limitations of the group.
4. Develop a set of guiding rules for the group.
5. Create an atmosphere conducive to group synergism.
6. Identify the questions to be addressed in advance.
7. Plan to address only one topic per meeting.

For major decisions and long-term group activities, arrange for team training that allows the group to learn the decision rules and responsibilities together. The steps for nominal group technique are:

1. Silently generate ideas, in writing.
2. Record ideas without discussion.
3. Conduct group discussion for clarification of meaning, not argument.
4. Vote to establish the priority or rank of each item.
5. Discuss vote.
6. Cast final vote.

Interviews, Surveys, and Questionnaires

Interviews, surveys, and questionnaires are important information-gathering techniques. They also foster cooperative working relationships; they encourage direct participation and inputs into project decision-making processes; and they provide an opportunity for employees at the lower levels of an organization to contribute ideas and inputs for decision making. The greater the number of people involved in the interviews, surveys, and questionnaires, the more valid the final decision. The following guidelines are useful for conducting interviews, surveys, and questionnaires to collect data and information for project decisions:

1. Collect and organize background information and supporting documents on the items to be covered by the interview, survey, or questionnaire.
2. Outline the items to be covered and list the major questions to be asked.
3. Use a suitable medium of interaction and communication: telephone, fax, electronic mail, face-to-face, observation, meeting venue, poster, or memo.
4. Tell the respondent the purpose of the interview, survey, or questionnaire and indicate how long it would take.
5. Use open-ended questions that stimulate ideas from the respondents.
6. Minimize the use of "yes" or "no" type of questions.
7. Encourage expressive statements that indicate the respondent's views.
8. Use the who, what, where, when, why, and how approach to elicit specific information.
9. Thank the respondents for their participation.
10. Let the respondents know the outcome of the exercise.

Multivote

Multivoting is a series of votes used to arrive at a group decision. It can be used to assign priorities to a list of items, or it can be used at team meetings after a brainstorming session has generated a long list of items. Multivoting helps to reduce such long lists to a few items, usually three to five. The steps for multivoting are:

1. Take a first vote. Each person votes as many times as desired, but only once per item.
2. Circle the items receiving a relatively higher number of votes (i.e., majority vote) than the other items.

3. Take a second vote. Each person votes for a number of items equal to one-half the total number of items circled in step 2. Only one vote per item is permitted.
4. Repeat steps 2 and 3 until the list is reduced to three to five items depending on the needs of the group. It is not recommended to multivote down to only one item.
5. Perform further analysis of the items selected in step 4, if needed.

PERSONNEL MANAGEMENT

Positive personnel management and interactions are essential for project success. Effective personnel management can enhance team building and coordination. The following guidelines are offered for personnel management in a project environment.

1. Leadership style
   - Lead the team rather than manage the team.
   - Display a personality of self-confidence.
   - Establish self-concept of your job functions.
   - Engage in professional networking without being pushy.
   - Be discreet with personal discussions.
   - Perform a self-assessment of professional strengths.
   - Dress professionally without being flashy.
   - Be assertive without being autocratic.
   - Keep up with the developments in the technical field.
   - Work hard without overexerting.
   - Take positive initiative when others procrastinate.
2. Supervision
   - Delegate when appropriate.
   - Motivate subordinates with vigor and objective approach.
   - Set goals and prioritize them.
   - Develop objective performance-appraisal mechanisms.
   - Discipline as required promptly.
   - Don't over-manage.
   - Don't shy away from mentoring or being mentored.
   - Establish credibility and decisiveness.
   - Don't be intimidated by difficult employees.
   - Use empathy in decision-making processes.

3. Communication
    - Be professional in communication approaches.
    - Do homework about communication needs.
    - Contribute constructively to meaningful discussions.
    - Exhibit knowledge without being patronizing.
    - Convey ideas effectively to gain respect.
    - Cultivate good listening habits.
    - Incorporate charisma into communication approaches.
4. Handling conflicts
    - Learn the politics and policies of the organization.
    - Align project goals with organizational goals.
    - Overcome fear of confrontation.
    - Form a mediating liaison between peers, subordinates, and superiors.
    - Control emotions in tense situations.
    - Don't take office conflicts home, and don't take home conflicts to work.
    - Avoid power struggles but claim functional rights.
    - Handle mistakes honestly without being condescending.

SYSTEMS INTEGRATION FOR ISO 9000

Systems integration permits sharing of resources. Physical equipment, concepts, information, and skills may be shared as resources. Systems integration is now a major concern of many organizations. Even some of the organizations that traditionally compete and typically shun cooperative efforts are beginning to appreciate the value of integrating their operations. For these reasons, systems integration has emerged as a major interest in business. System integration may involve physical integration of technical components, objective integration of operations, conceptual integration of management processes, or a combination of any of these.

Systems integration involves the linking of components to form subsystems and the linking of subsystems to form composite systems within a single department and/or across departments. It facilitates the coordination of technical and managerial efforts to enhance organizational functions, reduce cost, save energy, improve productivity, and increase the utilization of resources. Systems integration emphasizes the identification and coordination of the interface requirements between the components in an integrated system. The components and subsystems operate synergistically to optimize the performance of the total system. Systems integration ensures that all performance goals are satisfied with a minimum expenditure of time and resources. Integration can be achieved in several forms including the following:

1. *Dual-use integration:* This involves the use of a single component by separate subsystems to reduce both the initial cost and the operating cost during the project life cycle.
2. *Dynamic resource integration:* This involves integrating the resource flows of two normally separate subsystems so that the resource flow from one to or through the other minimizes the total resource requirements in a project.
3. *Restructuring of functions:* This involves the restructuring of functions and re-integration of subsystems to optimize costs when a new subsystem is introduced into the project environment.

System integration is particularly important when introducing new technology into an existing system. It involves coordinating new operations to coexist with existing operations, and it may require the adjustment of functions to permit sharing of resources, development of new policies to accommodate product integration, or realignment of managerial responsibilities. It can affect both hardware and software components of an organization. The following are guidelines and important questions relevant for systems integration:

- What are the unique characteristics of each component in the integrated system?
- How do the characteristics complement one another?
- What physical interfaces exist between the components?
- What data/information interfaces exist between the components?
- What ideological differences exist between the components?
- What are the data flow requirements for the components?
- Are there similar integrated systems operating elsewhere?
- What are the reporting requirements in the integrated system?
- Are there any hierarchical restrictions on the operations of the components of the integrated system?
- What are the internal and external factors expected to influence the integrated system?

- How can the performance of the integrated system be measured?
- What benefit/cost documentations are required for the integrated system?
- What is the cost of designing and implementing the integrated system?
- What are the relative priorities assigned to each component of the integrated system?
- What are the strengths of the integrated system?
- What are the weaknesses of the integrated system?
- What resources are needed to keep the integrated system operating satisfactorily?
- Which section of the organization will have primary responsibility for the operation of the integrated system?
- What are the quality specifications and requirements for the integrated system?

An integrated approach to ISO 9000 project management is presented in Figure 5-2. The process starts with a managerial analysis of the project effort. Goals and objectives are defined, a mission statement is written, and the statement of work is developed. After these, traditional project management approaches, such as selection of an organization structure, are employed. Conventional analytical tools including CPM and PERT are then mobilized. Optimization models are then used as appropriate. Some of the parameters to be optimized are cost, resource allocation, and schedule length. It should be understood that not all project parameters would be amenable to optimization. The use of commercial project management software should start only after the managerial functions have been completed. Some project management software has built-in capabilities for planning and optimization needs.

A frequent mistake in project management is the rush to use project management software without first completing the planning and analytical studies required by the project. Project management software should be used as a management tool, the same way a word processor is used as a writing tool. It will not be effective to start using the word processor without first organizing the thoughts about what is to be written. Project management is much more than just the project management software.

PROJECT BLUEPRINT

An outline of the functions to be carried out during a project should be made during the planning stage of the project. It may be necessary to rearrange the contents of the outline to fit the specific needs of an ISO 9000 project.

110 PROJECT BLUEPRINT

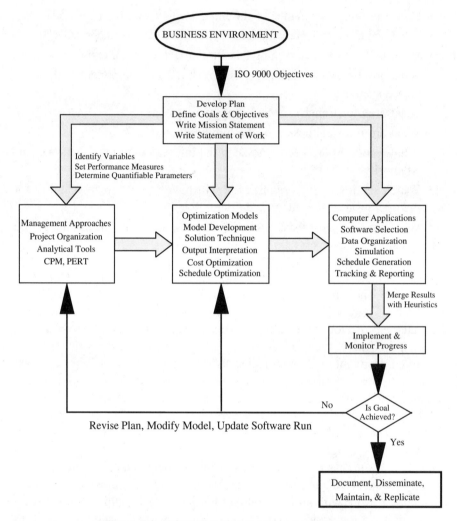

Figure 5-2. ISO 9000 Project Management Flowchart

Outline

Planning

  I. Specify project background
     A. Define current situation and process
        1. Understand the process
        2. Identify important variables
        3. Quantify variables

B. Identify areas for improvement
    1. List and discuss the areas
    2. Study potential strategy for solution
II. Define unique terminologies relevant to the project
    A. Industry-specific terminologies
    B. Company-specific terminologies
    C. Project-specific terminologies
III. Define project goal and objectives
    A. Write mission statement
    B. Solicit inputs and ideas from personnel
IV. Establish performance standards
    A. Schedule
    B. Performance
    C. Cost
V. Conduct formal project feasibility study
    A. Determine impact on cost
    B. Determine impact on organization
    C. Determine project deliverables
VI. Secure management support

Organizing

I. Identify project management team
    A. Specify project organization structure
        1. Matrix structure
        2. Formal and informal structures
        3. Justify structure
    B. Specify departments involved and key personnel
        1. Purchasing
        2. Materials management
        3. Engineering, design, manufacturing, etc.
    C. Define project management responsibilities
        1. Select project manager
        2. Write project charter
        3. Establish project policies and procedures
II. Implement triple C model
    A. Communication

1. *Determine communication interfaces*
2. *Develop communication matrix*
B. *Cooperation*
   1. *Outline cooperation requirements*
C. *Coordination*
   1. *Develop work breakdown structure*
   2. *Assign task responsibilities*
   3. *Develop responsibility chart*

Scheduling and Resource Allocation

I. *Develop master schedule*
   A. *Estimate task duration*
   B. *Identify task precedence requirements*
      1. *Technical precedence*
      2. *Resource-imposed precedence*
      3. *Procedural precedence*
   C. *Use analytical models*
      1. *CPM*
      2. *PERT*
      3. *Gantt chart*
      4. *Optimization models*

Tracking, Reporting, and Control

I. *Establish guidelines for tracking, reporting, and control*
   A. *Define data requirements*
      1. *Data categories*
      2. *Data characterization*
      3. *Measurement scales*
   B. *Develop data documentation*
      1. *Data update requirements*
      2. *Data quality control*
      3. *Establish data security measures*
II. *Categorize control points*
   A. *Schedule audit*
      1. *Activity network and Gantt charts*
      2. *Milestones*

      3. *Delivery schedule*
   B. *Performance audit*
      1. *Employee performance*
      2. *Product quality*
   C. *Cost audit*
      1. *Cost containment measures*
      2. *Percent completion versus budget depletion*
III. *Identify implementation process*
   A. *Comparison with targeted schedules*
   B. *Corrective course of action*
      1. *Rescheduling*
      2. *Re-allocation of resources*
IV. *Terminate the project*
   A. *Performance review*
   B. *Strategy for follow-up projects*
   C. *Personnel retention and releases*
V. *Document project and submit final report*

6
ISO 9000 PLANNING GUIDELINES

This chapter presents practical guidelines for preparing ISO 9000 activities. Preparation should be done throughout the organization. No cog-in-the-wheel process should be permitted to impede the enthusiasm for registration.

PHASES OF PREPARATION

The preparation can be divided into three major phases presented earlier in Chapter 2: *Planning phase, compliance phase, and registration phase.* The phases contain the elements listed below:

Planning Phase

1. Make the decision to comply with ISO 9000.
2. Undertake the education and training needed for general awareness.
3. Determine the scope of compliance (where, when, who).

Compliance Phase

1. Assess the current system.
2. Address deficiencies in the current processes.
3. Document existing quality system.
4. Establish internal audit function.

Registration Phase

1. Make decision about third-party registration.

2. Select a registrar.
3. Carry out registration process with full commitment.
4. Strive to maintain registration once it is achieved.

It is not enough to just achieve certification. It must be maintained.

STRATEGIC PLANNING

The key to a successful project is good planning. Project planning provides the basis for the initiation, implementation, and termination of a project. It sets guidelines for specific project objectives, project structure, tasks, milestones, personnel, cost, equipment, performance, and problem resolutions. The following common adages are applicable to strategic planning for ISO 9000:

- Plan your work, work your plan.
- Say what you do, do what you say.
- If it is not written down, then it did not happen.

An analysis of what is needed and what is available should be conducted in the planning phase of new projects. The availability of technical expertise within the organization and outside the organization should be reviewed. If subcontracting is needed, the nature of the contract should undergo a thorough analysis. The question of whether or not the project is needed at all should be addressed. The "make," "buy," "lease," "subcontract," or "do nothing" alternatives should be compared as a part of the planning process. Guidelines for project plans are as follows:

1. Use project plans to coordinate and integrate efforts.
2. Make effective use of available resources.
3. Be prepared to revise project plans when necessary.
4. Empower workers to estimate their own work.
5. Establish value-added tasks rather than routine activities.
6. Define explicit, specific, and tangible milestones.

In the initial stage of project planning, the internal and external factors that influence the project should be determined and given priority weights. Examples of internal influences on project plans include

- Infrastructure
- Project scope

- Labor relations
- Project location
- Project leadership
- Organizational goal
- Management approach
- Technical manpower supply
- Resource and capital availability

In addition to internal factors, a project plan can be influenced by external factors. An external factor may be the sole instigator of a project or it may manifest itself in combination with other external and internal factors. Such external factors include

- Public needs
- Market needs
- National goals
- Industry stability
- State of technology
- Industrial competition
- Government regulations

TIME-COST-PERFORMANCE CRITERIA

Project goals determine the nature of project planning. Project goals may be specified in terms of terms of time (schedule), cost (resources), or performance (output). A project can be simple or complex. Whereas simple projects may not require the whole array of project management tools, complex projects may not be successful without all of them. Project management techniques are applicable to a wide collection of problems ranging from manufacturing to medical services.

The techniques of project management can help achieve goals relating to better product quality, improved resource utilization, better customer relations, higher productivity, and fulfillment of due dates. These can be expressed in terms of the following project constraints:

- Performance specification
- Schedule requirements
- Cost limitations

118 TIME-COST-PERFORMANCE CRITERIA

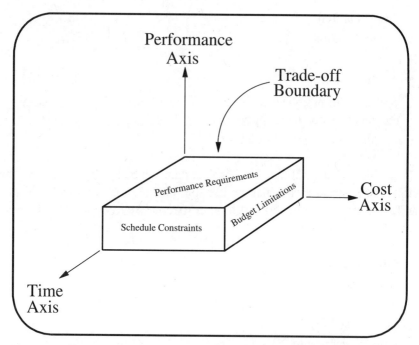

Figure 6-1. Performance-Time-Resource Trade-off Boundary

Time, cost, and performance form the basis for the operating characteristics of a business. These factors help to determine the basis for a successful project. Project control is the process of reducing the deviation between actual performance and planned performance. To control a project, we must be able to measure performance with respect to time and cost constraints.

Performance may be expressed in terms of quality, productivity, or any other measure of interest. Cost may be expressed in terms of resource expenditure or budget requirements. Schedule is typically expressed in terms of the timing of activities and the expected project duration. It is impossible to achieve an optimal simultaneous satisfaction of all three constraints. Consequently, it becomes necessary to compromise one constraint in favor of another. Figure 6-1 presents a model of the trade-offs among performance, time and cost.

Better performance can be achieved if more time and resources are available. If lower costs and tighter schedules are desired, then performance may have to be compromised and vice versa. From the point of view of the project manager, the project should be at the highest point along the performance axis. Of course, this represents an extreme case of getting something for nothing. From the point of view of the personnel, the project should be at the point indicating highest performance, longest time, and greatest resource availability. This, of course, may be an unrealistic expectation because time

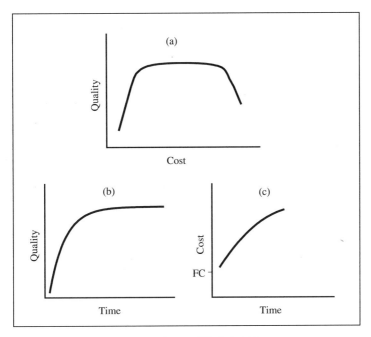

Figure 6-2. Trade-off Relationships

and resources are typically in short supply. A feasible trade-off strategy must be developed.

If we consider only two of the three constraints at a time, we can better study their respective relationships. Figure 6-2 shows some potential two-factor relationships. In plot (a), performance is modeled as the dependent variable, while cost is the independent variable. Performance increases as cost increases up to a point where performance levels off. If cost is allowed to continue to increase, performance eventually starts to drop. In plot (b), performance is modeled as being dependent on time. The more time that is allowed for a project, the higher the expected performance up to a point where performance levels off. In plot (c), cost depends on time. As project duration increases, cost increases. The increases in cost may be composed of labor cost, raw material cost, and/or cost associated with decreasing productivity. Note that there may be a fixed cost associated with a project even when a time schedule is not in effect.

Planning determines the nature of actions and responsibilities needed to achieve ISO 9000 objectives. It entails the development of alternate courses of action and the selection of the best action to achieve the objectives making up the goal. Planning determines what needs to be done, by whom, for whom, and when. Whether it is done for long-range (strategic) purposes or short-range (operational) purposes, planning should be one of the first steps of ISO 9000 project management.

ISO 9000 decisions will involve numerous personnel within the organization with various types and levels of expertise. In addition to the conventional roles of the project manager, specialized roles may be developed within the project scope. Such roles include the following:

1. *Technical specialist:* This person will have responsibility for addressing specific technical requirements of the project. In a large project, there may be several technical specialists working together to solve problems.
2. *Operations integrator:* This person will be responsible for making sure that all operational components of the effort interface correctly to satisfy project goals. This person should have good technical awareness and excellent interpersonal skills.
3. *Project specialist:* This person has specific expertise related to the specific goals and requirements of the ISO 9000 project. Even though a technical specialist may also serve as a project specialist, the two roles should be distinguished.

COMPONENTS OF AN ISO 9000 PLAN

Planning is an ongoing process that is conducted throughout the project life cycle. Initial planning may relate to overall organizational efforts. This is where specific projects to be undertaken are determined. Subsequent planning may relate to specific objectives of the selected project. In general, a project plan should consist of the following components:

1. *Executive summary of the plan:* This is a brief description of what is planned. Project scope and objectives should be enumerated. The critical constraints on the project should be outlined. The types of resources required and available should be specified. The summary should include a statement of how the project complements organizational and national goals, budget size, and milestones.
2. *Objectives:* The objectives should be very detailed in outlining what the project is expected to achieve and how the expected achievements will contribute to overall goals of a project. The performance measures for evaluating the achievement of the objectives should be specified.
3. *Approach:* The managerial and technical methodologies for implementing the project should be specified. The managerial approach

may relate to project organization, communication network, approval hierarchy, responsibility, and accountability. The technical approach may relate to company experience on previous projects and currently available technology.

4. *Policies and procedures:* Development of a project policy involves specifying the general guidelines for carrying out tasks within the project. Project procedure involves specifying the detailed method for implementing a given policy relative to the tasks needed to achieve the goal.

5. *Contractual requirements:* This portion of the project plan should outline reporting requirements, communication links, customer specifications, performance specifications, deadlines, the review process, project deliverables, delivery schedules, internal and external contacts, data security, policies, and procedures. This section should be as detailed as practically possible. Any item that has the slightest potential for creating problems later should be documented.

6. *Project schedule:* The project schedule signifies the commitment of resources against time in pursuit of objectives. A schedule should specify when the project will be initiated and when it is expected to be completed. The major phases of the project should be identified. The schedule should encompass reliable time estimates for tasks. The estimates may come from knowledgeable personnel, past records, or forecasting. Task milestones should be generated on the basis of objective analysis rather than arbitrary proclamations. The schedule in this planning stage constitutes the master schedule. Detailed activity schedules should be generated under specific functions.

7. *Resource requirements:* Resources, budget, and costs are to be documented in this section of the plan. Capital requirements should be specified by tasks. Resources may include personnel, equipment, and information. Special personnel skills, hiring, and training should be explained. Personnel requirements should be aligned with schedule requirements to ensure their availability when needed. Budget size and source should be presented. The basis for estimating budget requirements should be justified and the cost allocation and monitoring approach should be shown.

8. *Performance measures:* Measures of evaluating progress toward ISO 9000 objectives should be developed. The measures may be based on standard practices or customized needs. The method of monitoring, collecting, and analyzing the measures should also be specified. Corrective actions for specific problems should be outlined.

9. *Contingency plans:* Courses of action to be taken in case of undesir-

able events should be predetermined. Many projects have failed simply because no plans had been developed for emergency situations. In the excitement of getting a project under way, it is often easy to overlook the need for contingency plans.

10. *Tracking, reporting, and auditing:* These involve keeping track of the project plans, evaluating tasks, and scrutinizing the records of the project.

Not only must a plan be developed in accordance with the above guidelines, the plan must be widely disseminated. *An invisible plan is no plan.* Large project planning may include a statement about the feasibility of subcontracting part of the project work. Subcontracting may be needed for various reasons including lower cost, higher efficiency, or logistical convenience.

EMPLOYEE MOTIVATION

Motivation is an essential component of implementing project plans. National leaders, public employees, management staff, producers, and consumers may all need to be motivated about project plans that affect a wide spectrum of society. Those who will play active direct roles in the project must be motivated to ensure productive participation. Direct beneficiaries of the project must be motivated to make good use of the outputs of the project. Other groups must be motivated to play supporting roles to the project.

Various Approaches

Motivation may take several forms. For projects that are of a short-term nature, motivation could either be impaired or enhanced by the strategy employed. Impairment may occur if a participant views the project as a mere disruption of regular activities or as a job without long-term benefits. Long-term projects have the advantage of giving participants enough time to readjust to the project efforts.

Management involves getting things done through the efforts of people. An effective manager should be interested in both results and the people with whom he or she works. Management involves human aspects with behavioral and motivational implications. In order to get a worker to function effectively, he or she must be motivated. Some workers are inherently self-motivating, and there are workers for whom motivation is an external force that must be applied periodically. There are basically two types of workers: theory X workers and theory Y workers.

Theory X assumes that the worker is essentially uninterested and unmotivated to perform his or her work. Motivation must be instilled into the worker by the adoption of external motivating agents. A theory X worker is inherently indolent and requires constant supervision and prodding to get him or her to perform. To motivate a theory X worker, a mixture of managerial actions may be needed. The actions must be used judiciously based on the prevailing circumstances. Examples of motivation approaches under theory X are

- Rewards to recognize improved effort
- Strict rules to constrain worker behavior
- Incentives to encourage better performance
- Threats to job security associated with performance failure

Theory Y assumes that the worker is naturally interested and motivated to perform his or her job. The worker views the job function positively and uses self-control and self-direction to pursue project goals. Under theory Y, management has the task of taking advantage of the worker's positive intuition so that his or her actions coincide with the objectives of the project. Thus, a theory Y manager attempts to use the worker's self-direction as the principal instrument for accomplishing work. In general, Theory Y facilitates

- Worker-designed job methodology
- Worker participation in decision making
- A cordial management-worker relationship
- Worker individualism within acceptable company limits

Hierarchy of Needs

The needs of ISO 9000 participants must be taken into consideration in the planning stages. Maslow's hierarchy of needs (1954) is helpful for this purpose. Human needs follow a hierarchy as outlined here:

1. *Physiological needs:* the basic things of life such as food, water, housing, and clothing. This is the level where access to money is most critical.
2. *Safety needs:* security, stability, and freedom from threat of physical harm. The fear of adverse environmental impact may inhibit project efforts.
3. *Social needs:* social approval, friends, love, affection, and association. For example, public service projects may bring about a better economic outlook that may enable individuals to be in a better position to meet their social needs.

4. *Esteem needs:* accomplishment, respect, recognition, attention, and appreciation. These needs are important not only at the individual level, but also at the national level.
5. *Self-actualization needs:* self-fulfillment and self-improvement. They also involve the availability of opportunity to grow professionally. Work improvement projects may lead to self-actualization opportunities for individuals to assert themselves socially and economically. Job achievement and professional recognition are two of the most important factors that lead to employee satisfaction and better motivation.

Hierarchical motivation implies that the particular motivation technique utilized for a given person should depend on where the person stands in the hierarchy of needs. For example, the needs for esteem take precedence over the physiological needs when the latter are relatively well-satisfied. Money, for example, cannot be expected to be a very successful motivational factor of an individual who is already on the fourth level of the hierarchy of needs.

Credit Sharing

Give credit when and where it is due. Giving credit to workers when appropriate is very important for motivation. When a department collectively achieves a goal, it is important that relevant recognition be given to individual contributors. The department head alone should not get all the credit.

Motivating Factors

The characteristics of a task itself can affect motivation. Such factors are referred to as the *hygiene factors* and *motivators*. Hygiene factors are the necessary but not sufficient conditions for a contented worker. The negative aspects of the factors may create disgruntled workers, whereas their positive aspects do not necessarily enhance worker satisfaction. Examples include

1. *Administrative policies:* Bad policies can lead to the discontent of the workers, whereas good policies are viewed as routine with no specific contribution to improving worker satisfaction.
2. *Supervision:* A bad supervisor can make a worker unhappy and less productive, whereas a good supervisor cannot necessarily improve worker performance.
3. *Working conditions:* Bad working conditions can enrage workers, but good working conditions do not automatically generate improved productivity.
4. *Salary:* Low salaries can make a worker unhappy, disruptive, and uncooperative, but a raise will not necessarily make him or her per-

form better. Although a raise in salary will not necessarily increase professionalism, a reduction in salary will most certainly have an adverse effect on morale.
5. *Personal life:* A miserable personal life can adversely affect worker performance, but a happy life does not imply that he or she will be a better worker.
6. *Interpersonal relationships:* Good peer, superior, and subordinate relationships are important to keep a worker happy and productive, but extraordinarily good relationships do not guarantee that he or she will be more productive.
7. *Social and professional status:* Low status can force a worker to perform at his "level," whereas high status does not imply that he or she will perform at a higher level.
8. *Security:* A safe environment may not motivate a worker to perform better, but an unsafe condition will certainly impede his or her productivity.

Motivators should be inherent in the work itself. If necessary, work should be redesigned to include inherent motivating factors. Some guidelines for incorporating motivators into jobs are presented here:

1. *Achievement:* The job design should give consideration to opportunity for worker achievement and avenues for setting personal goals to excel.
2. *Recognition:* The mechanism for recognizing superior performance should be incorporated into the job design. Opportunities for recognizing innovation should be built into the job.
3. *Work content:* The work content should be interesting enough to motivate and stimulate the creativity of the worker. The amount of work and the organization of the work should be designed to fit a worker's needs.
4. *Responsibility:* The worker should have some measure of responsibility for how his or her job is performed. Personal responsibility leads to accountability, which invariably yields better work performance.
5. *Professional growth:* The work should offer an opportunity for advancement so that the worker can set his or her own achievement level for professional growth within a project plan.

The basic philosophy is that work can and should be made more interesting in order to motivate workers to perform better.

FEASIBILITY STUDY

The feasibility of a project can be ascertained in terms of technical factors, economic factors, or both. A feasibility study is documented with a report showing the ramifications of the project.

Technical feasibility refers to the ability of the process to take advantage of the current state of technology in pursuing further improvement. The technical capability of the personnel as well as the capability of the available technology should be considered.

Managerial feasibility involves the capability of the infrastructure of a process to achieve and sustain process improvement. Management support, employee involvement, and commitment are key elements required to ascertain managerial feasibility.

Economic feasibility involves the capability of the proposed project to generate economic benefits. A benefit-cost analysis and a break-even analysis are important aspects of evaluating the economic feasibility of new industrial projects. The tangible and intangible aspects of a project should be translated into economic terms to facilitate a consistent basis for evaluation.

Financial feasibility should be distinguished from economic feasibility. Financial feasibility involves the capability of the project organization to raise the appropriate funds needed to implement the proposed project. Financing or sponsorship can be a major obstacle in large multi-party endeavors.

Cultural feasibility deals with the compatibility of the proposed project with the cultural makeup of the organization. In labor-intensive projects, planned functions must be integrated with the local cultural practices and beliefs.

Social feasibility addresses the influences that a proposed project may have on the social system in the organization. The prevailing social structure may be such that certain categories of workers may not have access to certain functions. The effect of the project on the social status of the participants must be assessed to ensure compatibility.

Safety feasibility is another important aspect that should be considered in project planning. This refers to an analysis of whether the project is capable of being implemented and operated safely with minimal adverse effects on the environment. Unfortunately, environmental impact assessment is often not adequately addressed in complex projects.

A **politically feasible** project may be referred to as a "politically correct project." Political considerations often dictate the direction for a proposed project. This is particularly true for large projects with national visibility that may have significant government involvement and political implications.

In general terms, the elements of a feasibility analysis for an ISO 9000 project should cover the following:

1. *Need analysis:* This indicates recognition of a need for the project. The need may affect the organization itself, another organization, the market, or the public. A preliminary study is conducted to confirm and evaluate the need. A proposal of how the need may be satisfied is then made. Pertinent questions that should be asked include
    - Is the need significant enough to justify the proposed project?
    - Will the need still exist by the time the project is completed?
    - What are the alternate means of satisfying the need?
    - What are the economic, social, and environmental impacts of the need?
2. *Process work:* This is the preliminary analysis done to determine what will be required to satisfy the need. The work may be performed by a consultant who is an expert in the field. The preliminary study often involves system models or prototypes. A simulation of the proposed system can be carried out to predict the outcome before the actual project starts.
3. *Engineering or methodology:* This involves a detailed technical study of the project. Technology capabilities are evaluated as needed. Product design, if needed, should be done at this stage.
4. *Cost estimate:* This involves estimating project cost to an acceptable level of accuracy. Levels of around minus 5 percent to plus 15 percent are common at this stage of a project plan. Both initial and operating costs are included in the cost estimation. Estimates of capital investment, recurring, and nonrecurring costs should also be contained in the cost estimate document. Sensitivity analysis can be carried out on the estimated cost values to see how responsive the plan is to the estimated cost values.
5. *Financial analysis:* This involves an analysis of the cash-flow profile of the project. The analysis should consider rates of return, inflation, sources of capital, payback periods, break-even point, residual values, and sensitivity. This is a critical analysis because it determines whether or not and when funds will be available to the project. The project cash-flow profile helps to support the economic and financial feasibility of the project.
6. *Project impacts:* This portion of the feasibility study provides an assessment of the impact of the proposed project. Environmental, social, cultural, political, and economic impacts may be some of the factors that will determine how a project is perceived by the public. The value-added potential of the project should also be assessed.
7. *Conclusions and recommendations:* The feasibility study should end

with the overall outcome of the analysis. This may indicate an endorsement or disapproval of the project. Recommendations on what should be done should be included in this section of the feasibility study report.

BUDGET PLANNING

After the planning for a project has been completed, the next step is the allocation of the resources required to implement the project plan. This is referred to as budgeting of capital rationing. Budgeting is the process of allocating scarce resources to the various endeavors of an organization. It involves the selection of a preferred subset of a set of acceptable projects because of overall budget constraints. Budget constraints may result from restrictions on capital expenditures, shortage of skilled manpower, shortage of materials, or mutually exclusive projects. The budgeting approach can be used to express the overall organizational policy. The budget serves many useful purposes including

- Performance measure
- Incentive for efficiency
- Project selection criterion
- Expression of organizational policy
- Plan of how resources are expended
- Catalyst for productivity improvement
- Control basis for managers and administrators
- Standardization of operations within a given horizon

The preliminary effort in the preparation of a budget is the collection and proper organization of relevant data. The preparation of a budget for a project is more difficult than the preparation of budgets for regular and permanent organizational endeavors. Recurring endeavors usually generate historical data that serve as inputs to subsequent estimating functions. Projects, on the other hand, are often one-time undertakings without the benefits of prior data. The input data for the budgeting process may include inflationary trends, cost of capital, standard cost guides, past records, and forecast projections. Budget data collection may be accomplished by top-down budgeting or bottom-up budgeting.

Top-Down Budgeting

This involves collecting data from upper-level sources such as top and middle managers. The cost estimates supplied by the managers may come from judgments, past experiences, or past data on similar project activities. The cost estimates are passed to lower-level managers, who then break the estimates down into specific work components within the project. These estimates may, in turn, be given to line managers, supervisors, and so on to continue the process. At the end, individual activity costs are developed. The top management issues the global budget while the line worker generates specific activity budget requirements.

One advantage of the top-down budgeting approach is that individual work elements need not be identified prior to approving the overall project budget. Another advantage of the approach is that the aggregate or overall project budget can be reasonably accurate even though specific activity costs may contain substantial errors. There is, consequently, a keen competition among lower-level managers to get the biggest slice of the budget pie.

Bottom-Up Budgeting

This approach is the reverse of top-down budgeting. In this method, elemental activities, their schedules, descriptions, and labor skill requirements are used to construct detailed budget requests. The line workers that are actually performing the activities are requested to furnish cost estimates. Estimates are made for each activity in terms of labor time, materials, and machine time. The estimates are then converted to dollar values. The dollar estimates are combined into composite budgets at each successive level up the budgeting hierarchy. If estimate discrepancies develop, they can be resolved through intervention to senior management, junior management, functional managers, project managers, accountants, or financial consultants. Analytical tools such as learning-curve analysis, work sampling, and statistical estimation may be used in the budgeting process as appropriate to improve the quality of cost estimates.

All component costs and departmental budgets are combined into an overall budget and sent to top management for approval. A common problem with bottom-up budgeting is that individuals tend to overstate their needs, thinking that top management may cut the budget by some percentage. It should be noted, however, that sending erroneous and misleading estimates will only lead to a loss of credibility. Properly documented and justified budget requests are often spared the budget ax. Honesty and accuracy are invariably the best policies for budgeting.

WORK BREAKDOWN STRUCTURE

Project breakdown structure (PBS) refers to the breakdown of a project for planning, scheduling, and control purposes. The breakdown is often referred

WORK BREAKDOWN STRUCTURE

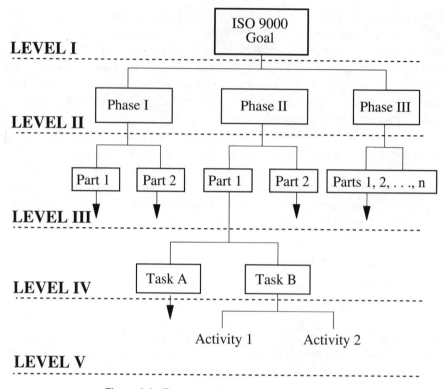

Figure 6-3. Example of Work Breakdown Structure

to as a work breakdown structure (WBS). This represents a hierarchy of tasks required to accomplish project objectives. Tasks that are contained in the WBS collectively describe the overall project. The tasks may involve physical products, services, and so on. The WBS serves to describe the link between the end objective and the tasks required to reach that objective. It shows work elements in the conceptual framework for planning. The objective of developing a WBS is to study the elemental components of a project in detail. Overall planning can be improved by using a WBS approach. A large project may be broken down into smaller sub-projects that may, in turn, be broken down into task groups.

Individual components in a WBS are referred to as WBS elements, and the hierarchy of each is designated by a level identifier. Elements at the same level of subdivision are said to be on the same WBS level. Descending levels provide increasingly detailed definition of project tasks. The complexity of a project and the degree of control desired determine the number of levels in the WBS. An example of a WBS is shown in Figure 6-3.

Each WBS component is successively broken down into smaller details at lower levels. The process may continue until specific project activities are reached. The basic approach for preparing a WBS is as follows:

1. Level 1 contains only the final project purpose. This item should be identifiable directly as an organizational budget item.
2. Level 2 contains the major subsections of the project. These subsections are usually identified by their contiguous location or by their related purposes.
3. Level 3 contains definable components of the level 2 subsections.

Subsequent levels are constructed in more specific details depending on the level of control desired. If a complete WBS becomes too crowded, separate WBSs may be drawn for the level 2 components. A specification of work (SOW) or WBS summary should normally accompany the WBS. A SOW is a narrative of the work to be done. It should include the objectives of the work, its nature, its resource requirements, and a tentative schedule. Each WBS element is assigned a code that is used for its identification throughout the project life cycle. Alphanumeric codes may be used to indicate element level as well as component group.

INFORMATION FLOW FOR ISO 9000

Information flow is very crucial in project planning. Information is the fuel for decisions. The value of information is measured in terms of the quality of the decisions that can be obtained from the information. What appears to be valuable information to one user may be useless to another. Similarly, the timing of the information can significantly affect its decision-making utility. The same information that is useful in one instant may be useless in another. Some of the crucial factors affecting the usefulness of information are

- Accuracy
- Timeliness
- Relevance
- Reliability
- Validity
- Completeness
- Clearness
- Comprehensibility

Proper information flow in project management ensures that tasks are

132   INFORMATION FLOW FOR ISO 9000

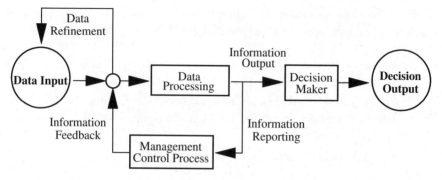

Figure 6-4. Information Flow for ISO 9000

accomplished when, where, and how they are needed. Figure 6-4 presents an information flow model for ISO 9000 planning.

Information starts with raw data (e.g., numbers, facts, specifications). The data may pertain to raw material, technical skills, or other factors relevant to the goal, and are processed to generate information in the desired form. The information feedback model acts as a management control process that monitors project status and generates appropriate control actions. The information is processed through an information feedback loop, and the feedback information is used to enhance the future processing of input data to generate additional information. The final information output provides the basis for improved management decisions. The key questions to ask when requesting, generating, or evaluating information for project management are

- What data are needed to generate the information?
- Where are the data going to come from?
- When will the data be available?
- Is the data source reliable?
- Are there enough data?
- Who needs the information?
- When is the information needed?
- In what form is the information needed?
- Is the information relevant to project goals?
- Is the information accurate, clear, and timely?

As an example, the information flow model previously described may be implemented to facilitate the inflow and outflow of information linking several functional areas involved in an ISO 9000 effort, such as the design department, the manufacturing department, the marketing department, and the customer relations department. The lack of communication between functional departments has been blamed for much of the organizational problems in industry. The use of a standard information flow model can help alleviate many of such communication problems.

Too much information is as bad as too little information. It can impede the progress of a project. The marginal benefit of information decreases as its size increases. However, the marginal cost of obtaining additional information may increase as the size of the information increases.

The optimum size of information is determined by the point that represents the widest positive difference between the value of information and its cost. The costs associated with information can often be measured accurately. However, the benefits may be more difficult to document. The size of information may be measured in terms of a number of variables including the number of pages of documentation, the number of lines of code, and the amount of computer storage needed. The amount of information presented for project management purposes should be condensed to cover only what is needed to make effective decisions. Information condensation may involve pruning the information that is already available or limiting what needs to be acquired.

The cost of information is composed of the cost of the resources required to generate the information. The required resources may include computer time, personnel hours, software cost, and so on. Unlike the value of information, which may be difficult to evaluate, the cost of information is somewhat more tractable. However, the development of accurate cost estimates prior to actual project execution is not trivial. The degree of accuracy required in estimating the cost of information depends on the intended uses of the estimates. Cost estimates may be used as general information for control purposes; they may also be used as general guides for planning purposes or for developing standards. The bottom-up cost-estimation approach is a popular method that involves breaking the cost structure down into its component parts. The cost of each element is then established. The sum of these individual cost elements gives an estimate of the total cost of the information.

Investments for information acquisition should be evaluated just like any other capital investment. The value of information is determined by how the information is used. In project management, information has two major uses. The first use relates to the need for information to run the daily operations of a project. Resource allocation, material procurement, replanning, rescheduling, hiring, and training are just a few of the daily functions for which information is needed. The second major use relates to the need for information to

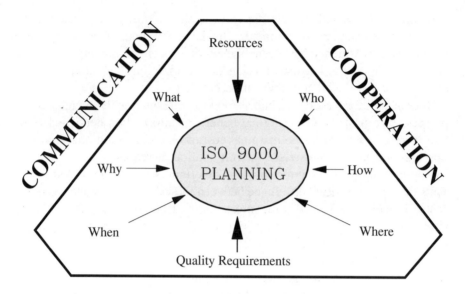

Figure 6-5. Triple C Model for ISO 9000

make long-range decisions. The value of information for such long-range decision making is even more difficult to estimate because the future cost of not having the information today is unknown.

The classical approach to determining the value of information is based on the value of perfect information. The expected value of perfect information is the maximum expected loss caused by imperfect information. Using probability analysis or other appropriate quantitative methods, the project analyst can predict what a project outcome might be if certain information is available or not available.

TRIPLE C APPROACH TO PLANNING

Communication is the prerequisite for cooperation. The Triple C concept of project management may be applied to ensure that *communication, cooperation,* and *coordination* functions are carried out to facilitate total quality management. The Triple C model encourages employee involvement and customer participation. Figure 6-5 presents a pictorial representation of the Triple C model.

Communication

Communication is the very complex process of sending, receiving, and understanding messages. Every functional group or individual that can influence quality throughout the hierarchies of the organization should be informed and made aware of the quality objectives. Specific details of communication should include

- The quality objective
- The scope of the quality objective
- The plan for achieving the objective
- The desired personnel interfaces
- The direct and indirect benefits of higher quality
- The expected cost of low quality
- The desired personnel contribution to the quality objective

The communication channel must be kept open throughout the quality-improvement effort. In addition to in-house communication, external communication needs should also be addressed. To facilitate communication, management must

- Exude commitment to quality improvement
- Endorse the communication responsibility matrix
- Facilitate multichannel communication interfaces
- Support internal and external communication needs
- Help resolve organizational and communication conflicts
- Promote both formal and informal communication links

When clear communication is maintained between management and employees and between peers, many project problems can be averted. Project communication may be carried out in one or more of the following formats:

- One-to-many
- One-to-one
- Many-to-one
- Written and formal
- Written and informal
- Oral and formal

- Oral and informal
- Nonverbal gesture

Good communication is effected when what is implied is perceived as intended. Effective communications are vital to the success of any project. Despite the awareness that proper communications form the blueprint for project success, many organizations still fail in their communication functions. Factors that influence the effectiveness of communication within a project organization structure include the following:

1. *Personal perception:* Each person perceives events on the basis of personal psychological, social, cultural, and experiential backgrounds. As a result, no two people can interpret a given event the same way. The nature of events is not always the critical aspect of a problem situation. Rather, the problem is often the different perceptions of the various people involved.
2. *Psychological profile:* The psychological makeup of each person determines personal reactions to events or words. Thus, individual needs and levels of thinking will dictate how a message is interpreted.
3. *Social environment:* Communication problems sometimes arise because people have been conditioned by their prevailing social environments to interpret certain things in unique ways. Vocabulary, idioms, organizational status, social stereotypes, and economic situation are among the social factors that can thwart effective communication.
4. *Cultural background:* Cultural differences are among the most pervasive barriers to project communications, especially in today's multicultural organizations. Language and cultural idiosyncrasies often determine how communication is approached and interpreted.
5. *Semantic and syntactic factors:* Semantic and syntactic barriers to communications usually occur in written documents. Semantic factors are those that relate to the intrinsic knowledge of the subject of the communication. Syntactic factors are those that relate to the form in which the communication is presented. The problems created by these factors become acute in situations where response, feedback, or reaction to the communication cannot be observed.
6. *Organizational structure:* Frequently, the organization structure in which a project is conducted has a direct influence on the flow of information and, consequently, on the effectiveness of communication. Organization hierarchy may determine how different personnel levels perceive a given communication.
7. *Communication media:* The method of transmitting a message may

also affect the value ascribed to the message and, consequently, how it is interpreted or used. Note that once a message is communicated and received, whether intentional or not, it is impossible to disregard.
8. *Body language:* We communicate with all of our senses, some of them subconsciously. Facial expressions and body movements, even when unintended, can influence communication and the perceptions of those involved.

The common barriers to project communication are

- Inattentiveness
- Lack of organization
- Outstanding grudges
- Preconceived notions and personal bias
- Ambiguous presentation
- Emotions and sentiments
- Lack of communication feedback
- Sloppy and unprofessional presentation
- Lack of confidence in the communicator
- Lack of confidence by the communicator
- Low credibility of communicator
- Unnecessary technical jargon
- Too many people involved
- Untimely communication
- Arrogance or imposition
- Lack of focus

The following recommendations for improving the effectiveness of communication may be implemented as appropriate for any of the forms of communication listed earlier. The suggestions apply to both the communicator and the audience.

1. Never assure that the integrity of the information sent will be preserved as the information passes through several communications channels. Information is generally filtered, condensed, or expanded by receivers before they relay it to the next destination. When preparing communication that needs to pass through several organization structures, one safeguard is to compose the original information in a concise form to minimize the need for recomposition.
2. Give the audience a central role in the discussion. Such participation

can help make people feel part of the project effort and responsible for the project's success. They can then have a more constructive view of project communication.
3. Do homework and think through the intended accomplishment of the communication. This helps eliminate trivial and inconsequential communication efforts.
4. Carefully plan the organization of the ideas embodied in the communication. Use indexing or points of reference whenever possible. Grouping ideas into related chunks of information can be particularly effective. Present the short messages first. Short messages help create focus, maintain interest, and prepare people for the longer messages to follow.
5. Highlight why the communication is of interest and how it is intended to be used. Full attention should be given to the content of the message with regard to the prevailing project situation.
6. Elicit the support of those around you by integrating their ideas into the communication. The more people feel they have contributed to the issue, the more expeditious they are in soliciting the cooperation of others. This can quickly garner support for the communication purpose.
7. Be responsive to the feelings of others. It takes two to communicate. Anticipate and appreciate the reactions of audience members. Recognize their operational circumstances and present your message in the form they can relate to.
8. Accept constructive criticism. Nobody is infallible. Use criticism as a springboard to higher communication performance.
9. Exhibit interest in the issue in order to arouse the interest of your audience. Avoid delivering your messages as a matter of routine organizational requirements.
10. Obtain and furnish feedback promptly. Clarify vague points with examples.
11. Communicate at the appropriate time, at the right place, to the right people.
12. Reinforce words with positive action. Never promise what cannot be delivered. Value your own credibility.
13. Maintain eye contact during oral communication and read the facial expressions of your audience to obtain real-time feedback.
14. Concentrate on listening as much as speaking. Evaluate both the implicit and explicit meanings of statements.
15. Document communication transactions for future references.
16. Avoid asking questions that can be answered with "yes" or "no."

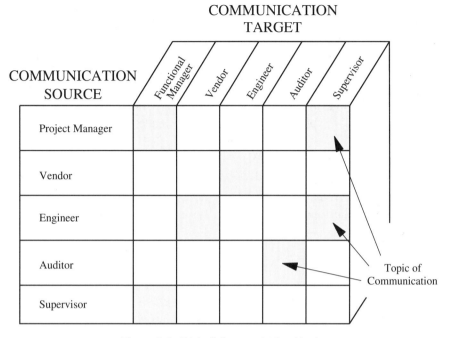

Figure 6-6. Triple C Communication Matrix

Use relevant questions to focus the attention of the audience. Use questions that make people consider their words: "How do you think this will work?" instead of "Do you think this will work?"

17. Avoid patronizing the audience. Respect their judgment and knowledge.
18. Speak and write in a controlled tempo. Avoid emotionally charged voice inflections.
19. Create an atmosphere for formal and informal exchange of ideas.
20. Summarize the objectives of the communication and how they will be achieved.

Figure 6-6 shows an example of a communication responsibility matrix, which shows the linking of sources of communication and targets of communication. Cells within the matrix indicate the subject of the desired communication. There should be at least one filled cell in each row and each column of the matrix. This assures that each individual of each department has at least one communication source or target associated with him or her. With a communication responsibility matrix, a clear understanding of what needs to be communicated to whom can be developed.

Cooperation

Not only must the workforce be informed and educated about quality, but their cooperation must also be explicitly sought. Merely giving a nod to the quality effort may not be an indication of full cooperation. The justification for the quality objective must be explained to all the groups concerned including the union, managers, clients, and suppliers. To seek the cooperation of employees, explicit statements must be made about

- What cooperative efforts are needed
- The importance of cooperation to the quality program
- Rewards of cooperation for quality
- Organizational impact of cooperation
- Implications of low quality

A documentation of the prevailing level of cooperation is useful for winning further support for quality improvement. Clarification of quality priorities will facilitate personnel cooperation. Relative priorities of multiple projects should be specified so that a quality improvement project that is of high priority to the organization will also be of high priority to all groups within the organization.

More projects fail because of a lack of cooperation and commitment than any other project factors. To secure and retain the cooperation of project participants, their first reaction to the project must be positive. The most positive aspects of a project should be the first items of project communication. For project management, the following different types of cooperation should be understood.

Functional Cooperation
This is cooperation induced by the nature of the functional relationship between two groups that may be required to perform related functions that can only be accomplished through mutual cooperation.

Social Cooperation
This is the type of cooperation effected by the social relationship between two groups. The prevailing social relationship motivates cooperation that may be useful in getting project work done.

Legal Cooperation
Legal cooperation is the type of cooperation that is imposed through some authoritative requirement. In this case, the participants may have no choice other than to cooperate.

Administrative Cooperation
This is cooperation brought on by administrative requirements that make it imperative that two groups work together on a common goal.

Associative Cooperation
This is a type of cooperation that may also be referred to as collegiality. The level of cooperation is determined by the association that exists between two groups.

Proximity Cooperation
Cooperation caused by geographical closeness is referred to as proximity cooperation. Being close makes it imperative that the two groups work together.

Dependency Cooperation
This is cooperation caused by the fact that one group depends on another group for something important. Such dependency is usually of a two-way nature.

Imposed Cooperation
In this type of cooperation, external agents must be employed to induce cooperation between two groups. This is applicable for cases where the two groups have no natural reason to cooperate. The approaches presented earlier for seeking cooperation can become very useful in this type of situation.

Lateral Cooperation
Lateral cooperation involves cooperation with peers and immediate associates; it is often easy to achieve because existing lateral relationships create a conducive environment for project cooperation.

Vertical Cooperation
Vertical or hierarchical cooperation refers to cooperation that is implied by the hierarchical structure of the project. For example, subordinates are expected to cooperate with their superiors.

Whichever type of cooperation is available in a project environment, the cooperative forces should be channeled toward achieving project goals. Some guidelines for securing cooperation for most projects are as follows:

- Establish achievable goals for the project.
- Clearly outline the individual commitments required.
- Integrate project priorities with existing priorities.
- Eliminate the fear of job loss that is due to industrialization.

Table 6-1 Responsibility Matrix for Quality Improvement

	Actors				
Responsibilities	Management	Marketing	Manufacturing	Person A	Person B
1. Survey of Customer Needs	I	R	I	C	S
2. Funding for Quality Improvement	R	I	I	I	I
3. Design Review	S	C	R	R	I
4. Monitor Progress	R	R	R	R	R
5. Institutionalize Improvement	R	R	R	R	R

Responsibility Codes:
 R: Responsible
 C: Contribute
 I: Inform
 S: Support

- Anticipate and eliminate potential sources of conflict.
- Use an open-door policy to address project grievances.
- Remove skepticism by documenting the merits of the project.

Commitment

Cooperation must be supported with commitment. To cooperate is to support the ideas of a project. To commit is to willingly and actively participate in project efforts again and again through the thick and thin of the project. Provision of resources is one way that management can express commitment to a project. Using the Triple C model plus commitment equals ISO 9000 success.

Coordination

Having successfully initiated the communication and cooperation functions, the efforts of the groups and individuals (subsystems) in the organization must be carefully coordinated. Coordination facilitates harmonious and systematic integration of the contribution of each subsystem to overall quality objectives. The development of a responsibility matrix may be quite useful in this regard. Table 6-1 presents an example of a responsibility matrix for quality improvement. The matrix consists of columns of "actors" and rows of specific responsibilities or required actions. The actors can be individuals, groups of people, or functional departments.

Cells within the responsibility matrix are filled with relationship codes that indicate who is responsible for what and the nature of the responsibility. R indicates that the actor is responsible for the specified action. I indicates that the actor must be informed of what is going on with respect to the action. S indicates that the actor must support the efforts to carry out the action. C indicates that the individual must contribute to the stated action. The matrix can be as large as needed to cover the scope of the quality improvement effort. Because quality improvement is supposed to be everybody's responsibility, it is recommended that no cell in a responsibility matrix be left blank. Everybody should be, at least, informed of the ongoing efforts.

The responsibility matrix helps to avoid overlooking critical communication and functional requirements as well as obligations in the quality management effort. Every line of responsibility in the matrix must have at least one R code in the row, and every individual or department listed in the matrix must have at least one code in the column. Table 6-2 presents a more detailed responsibility matrix for ISO 9000. The R1 and R2 codes indicate shared responsibility, R1 for primary responsibility and R2 for secondary responsibility. The matrix can help resolve issues such as

- Functional responsibilities for specific quality functions
- Quality performance measurement standards
- Information transfer and feedback loop
- Who will do what
- Who will inform whom of what
- Whose approval is needed for what
- Who is responsible for which results
- What personnel interfaces are involved
- What support is needed from whom and when

Team Interaction

Team interaction is a participative method for implementing the systems approach to ISO 9000. A functional responsibility matrix facilitates team interaction, and highlights the responsibility, authority, and interrelation of all personnel who manage, perform, and verify ISO 9000 requirements. When combined with the Triple C approach, it offers the following benefits:

- Improved product quality
- Higher employee productivity

Table 6-2 ISO 9000 Responsibility Matrix

ISO 9000 Quality Systems Requirements	Mgt	QA	Mfg	Dsgn Eng	Doc Cntl	Pur	Matl	Fin	Pln	Acct	H/R	Mfg Serv	MIS	CIM
1. Mgmt Responsibility	R													
2. Quality System	R	C	C	C	C	C	C	C	C	C	C	C	C	C
3. Contract Review		C	C	C					C	R				
4. Design Control		C	C	R	C	C	C	C	C	C				
5. Document Control		C	C	C	R				C					
6. Purchasing		C	C	C		R								
7. Purchaser Supplied Product		R2	C	C			C			C				
8. Product Ident., Traceability		C					R1			C				
9. Process Control		C	C	C					C					C
10. Inspection & Testing		R1	C	C			C							
11. Inspection, Measuring & Test Equip.		C									C			
12. Inspection & Test Status		R2												
13. Control of Nonconforming Product		R1	C	C			C		C	C				
14. Corrective Action		R	C	C			C		C	C				
15. Handling, Storage, Packing & Delivery			R1	C		C	R2			C				
16. Quality Records		R2	R1								C		C	C
17. Internal Quality Audits		R	C											
18. Training			R2								C	R1		
19. Servicing		C	C							R				
20. Statistical Techniques		C	R	C						C				

Dsgn: Design Engineering   Doc Cntl: Document Control   Pur: Purchasing   Matl: Materials   Fin: Finance   Pln: Planning

- Increased job satisfaction
- Consistent platform for cooperation
- Employee development
- More visible constructive interaction between departments
- Increased awareness of organizational efforts to improve quality

Conflict Resolution

When implemented as an integrated process, the Triple C model can help avoid conflicts in ISO 9000 efforts. When conflicts do develop, it can help in resolving the conflicts. Several sources of conflicts can exist in large projects. Some of these are hereafter discussed.

Schedule Conflict

Conflicts can develop because of improper timing or sequencing of project tasks. This is particularly common in large, multiple projects. Procrastination can lead to having too much to do at once, thereby creating a clash of project functions and discord between project team members. Inaccurate estimates of time requirements may lead to unfeasible activity schedules. Project coordination can help avoid schedule conflicts.

Cost Conflict

Project cost may not be generally acceptable to the clients of a project. This will lead to project conflict. Even if the initial cost of the project is acceptable, a lack of cost control during project implementation can lead to conflicts. Poor budget allocation approaches and the lack of a financial feasibility study will cause cost conflicts later on in a project. Communication and coordination can help prevent most of the adverse effects of cost conflicts.

Performance Conflict

If clear performance requirements are not established, performance conflicts will develop. A lack of clearly defined performance standards can lead each person to evaluate his or her own performance based on personal value judgments. In order to uniformly evaluate quality of work and monitor project progress, performance standards should be established by using the Triple C approach.

Management Conflict

There must be a two-way alliance between management and the project team. The views of management should be understood by the team, and the views of the team should be appreciated by management. If this does not happen, management conflicts will develop. A lack of a two-way interaction can lead to strikes and industrial actions that can be detrimental to project objectives. The Triple C approach can help create a conducive environment for dialogue between management and the project team.

Technical Conflict
If the technical basis of a project is not sound, technical conflicts will develop. New industrial projects are particularly prone to technical conflicts because of their significant dependence on technology. The lack of a comprehensive technical feasibility study will lead to technical conflicts. Performance requirements and systems specifications can be integrated through the Triple C approach to avoid such conflicts.

Priority Conflict
Priority conflicts can develop if project objectives are not defined properly and applied uniformly across a project. The lack of project definition can lead each project member to define his or her own goals, which may be in conflict with the intended goal of a project. A project mission that is inconsistent can also be another potential source of priority conflicts. Over-assignment of responsibilities with no guidelines for priorities can also lead to conflicts. Communication can help defuse priority conflicts.

Resource Conflict
Resource allocation problems are a major source of conflict in project management. Competition for resources, including personnel, tools, hardware, software, and so on, can lead to disruptive clashes among project members. The Triple C approach can help secure resource cooperation.

Power Conflict
Project politics can lead to power plays that can adversely affect the progress of a project. Project authority and power should be clearly delineated. Authority is the control that a person has by virtue of his or her functional post; power relates to the clout and influence a person can exercise because of connections within the administrative structure. Popular people can often wield a lot of power in spite of low or nonexistent authority. The Triple C model can facilitate a positive marriage of authority and power to the benefit of project goals. This will help define clear leadership for a project.

Personality Conflict
Personality conflict is a common problem in projects involving a large group of people. The larger a project, the larger the size of the management team needed to keep things running. Unfortunately, the larger management team

creates an opportunity for personality conflicts. Communication and cooperation can help defuse personality conflicts.

In summary, conflict resolution through Triple C can be achieved by observing the following guidelines:

1. Confront the conflict and identify the underlying causes.
2. Be cooperative and receptive to negotiation as a mechanism for resolving conflicts.
3. Distinguish between proactive, inactive, and reactive behaviors in a conflict situation.
4. Use communication to defuse internal strife and competition.
5. Recognize that short-term compromise can lead to long-term gains.
6. Use coordination to work toward a unified goal.
7. Use communication and cooperation to turn a competitor into a collaborator.

7
QUALITY AUDIT TECHNIQUES

One of the fundamental principles of industrial engineering is the questioning and listening concept of organizational review. This principle is applicable to the implementation of internal audits for ISO 9000 process. Question everything and listen to all responses. It is the responsibility of management to assist the site ISO coordinator and documentation coordinators in locating the people responsible for the answers to internal audit questions. The questions presented in this chapter are based on the requirements of ISO 9001.

PURPOSE OF INTERNAL AUDIT

The purpose of an internal quality audit is to examine the effectiveness of management-directed control systems and programs within the organization. The philosophy of quality assurance is based on *prevention* rather than on detection of problems. But problems do occur and, and when they do, emphasis is placed on

- Early detection of the problem
- Evaluation of the scope of the problem
- Identification of the root cause of the problem
- Eradication of problem causes

The management of an organization should implement planning, organizing, and control strategies to

- Prevent problems
- Identify problems
- Solve problems
- Safeguard against reoccurrence of problems

Quality problems that are not addressed promptly will result in

- Customer dissatisfaction
- Loss of profit
- Loss of employee morale
- Disruption of work process

Quality audits are formal, systematic, and independent evaluations of certain aspects of an organization. The results of an audit are based on facts. The effectiveness and integrity of an audit depend heavily on the skills, tools, and training of the audit team. The questions presented in this chapter are intended to constitute a tool for the audit team. The purpose of audit should not be to find faults and criticize. Rather, it should be an objective review of a process and a presentation of constructive recommendations. Some of the specific goals of a quality audit include the following:

- Certify that products are fit for use.
- Certify that written procedures exist.
- Check adherence to legal or regulatory requirements.
- Identify product deficiencies.
- Ensure conformance to specification.
- Ensure that corrective actions are taken and are effective.
- Instill a sense of responsibility and accountability.

TYPES OF AUDITS

An audit is defined as a formal examination or verification of records. There are three major types of quality audit as hereafter outlined.

System Audit

A system audit is the largest and most extensive type of quality audit. A typical system audit can last from a few hours to several days. System audits scrutinize the interactive quality structure of the organization as a whole and the effect of the system on the product or the service.

The purpose of a system audit is to verify, through objective evidence, whether or not the quality management system and organizational plans are indeed carried out as specified in the documentation. System audits may be external (e.g., an audit of a supplier) or internal (e.g., an audit of a department's accounting system). The number of people required to perform the audit depends on the extent and intensity required to determine conformance. The third-party ISO registration agency conducts a systems audit.

Process Audit

A process audit looks at a portion of the quality management system. It concentrates on a specific system or organization to verify conformance to standards, methods, procedures, or other requirements. A systems audit can be viewed as a series of process audits. A company's internal audit team conducts process audits of the quality management system.

Product Audit

A product audit is an assessment of the final product or service and its "fitness for use" evaluated against the intent of the product or service. Product audits may be performed immediately prior to shipment, at various stages in the production cycle, or externally at the customer's end.

System, process, and product audits may be further broken down into one or two categories: internal and external audits.

Internal Audit
An internal audit is performed within an organization to measure its own performance, strengths, and weaknesses against its own established policies, procedures, and systems. It is also known as a first-party audit.

External Audit
An audit performed by an outside source is referred to as an external audit. If it is performed by a customer (e.g., for a precontract survey), it is called a second-party audit. If an outside agency performs the audit under contract, it is called a third-party audit. The roles of lead auditors and internal auditors are presented in Chapter 1. Third-party assessment steps are presented in Chapter 2.

MANAGEMENT REVIEW

The purpose of management review is to provide senior management a means for confirming the continuing suitability and effectiveness of all aspects of an organization's activities that affect the quality of the products and services.

The management representative is responsible for conducting this review. All senior management personnel are responsible for attending the management review meetings and representing their respective departments.

Management reviews are held weekly to assess portions of the quality system. Weekly topics are determined by the quality management staff. Semiannually, a management review can be held to assess the quality system in general. At these meetings, internal audit reports, customer complaints, corrective action plans and results, training requirements, and company objectives are reviewed against company policies and procedures. The two stages of assessment are as follows:

1. Determine if the documented quality system, as defined by the quality system manual and subsequent documentations, complies with the standard.
2. Determine if they are being followed in actual practice.

AUDIT TEAM PREPARATION

Prior to the audit, the team leader and auditors should prepare for the audit by

1. Reviewing the functional procedures of the organization being audited,
2. Reviewing the appropriate sections of the ISO 9001 standard and the applicable sections in the quality system manual,
3. Reviewing past internal audit reports and nonconformances for the area,
4. Creating an audit strategy (for example, focus on particular organizations, processes, procedures, or sections of the standard that have had problems in the past or have not been covered by a pervious audit),
5. Predetermining specific questions to ask, records to review, or areas to pursue, and
6. Setting a time for the audit and notifying the audit team and the auditee (for extensive audits, an opening meeting may be helpful to briefly cover the strategy and to assign people to various tasks).

SAMPLING STRATEGY

Since there is a limited time to cover all the items of interest during an audit, it is often necessary to select a limited sample to evaluate. Generally, a few samples are adequate to determine the status of the product or process. It is best to take samples at areas of product or process interface. When recording sample data, note the following items:

1. Location of samples (stage, process, department, etc.)
2. Part number and name
3. Product identification
4. Lot number or other identifiers
5. Time and date
6. Any other relevant observations

Proper sampling techniques should be used in collecting data. This is usually problem-dependent. There are several different types of data:

- Variables data
- Attributes data
- Countable data
- Rating data
- Ranking data

Data Measurement Scales

Identification of relevant measurement scales is important for the development of a data collection strategy for quality improvement. The common scales are as follows:

1. The *nominal scale* is the lowest level of measurement scales. It classifies items into categories. The categories are mutually exclusive and collectively exhaustive (i.e., the categories do not overlap and they cover all possible categories of the characteristics being observed). Gender and color are two examples of classifications on a nominal scale.
2. An *ordinal scale* is distinguished from a nominal scale by the property of order among the categories. An example is the process of prioritizing tasks for resource allocation. In quality control, the ABC classification of items based on the Pareto distribution is an example of a measurement on an ordinal scale.
3. An *interval scale* is distinguished from an ordinal scale by having equal intervals between the units of measure. The assignment of quality ratings ranging from 0 to 100 to a product is an example of a measurement on an interval scale. Temperature is a good example of an item that is measured on an interval scale. Even though there is a zero

point on the temperature scale, it is an arbitrary relative measure. Other examples of interval scales are IQ measurements and aptitude ratings.
4. A *ratio scale* has the same properties of an interval scale, but with a true zero point. For example, an estimate of a zero time unit for the duration of a task is a ratio scale measurement. Other examples of items measured on a ratio scale are volume, length, height, weight, and inventory level. Many of the items measured in a process and quality improvement program will be on a ratio scale.

In addition to the measurement scale, data can be classified based on their inherent nature. Examples of the relevant classifications are transient data, recurring data, static data, and dynamic data. *Transient data* are defined as volatile data that are encountered once during an expert system consultation and are not needed again. Transient data need not be stored in a permanent database unless they may be needed for future analysis or uses.

Recurring data are encountered frequently enough to necessitate storage on a permanent basis. Recurring data may be further categorized into *static data* and *dynamic data*. Static data will retain their original parameters and values each time they are encountered during an expert system consultation. Dynamic data have the potential for taking on different parameters and values each time they are encountered.

For proper data analysis, data should be recorded in such a way that the structure and format are easy to use and understand. This will enable the analyst to identify possible stratifications and populations that exist in the data if any. The data structure can be described as follows:

- Objective(s) of the data collection
- Quality characteristics of the measurement with clear operational definition
- Date of collection
- Operators taking the measurements
- Instruments or machines used
- All the processes used for samples
- All other sources of the data collection, including different sites, labs, method, material, etc.

Sample Characteristics

A sample is a subset of a population that is selected for observation and statistical analysis. Inferences are drawn about the population based on the analysis of the sample. The reasons for using sampling rather than complete population enumeration are as follows:

1. It is more economical to work with a sample.
2. There is a time advantage to using a sample.
3. Populations are typically too large to work with.
4. A sample is more accessible than the whole population.
5. In some cases, the sample may have to be destroyed during the analysis.

There are three primary types of samples. They differ in the manner in which their elementary units are chosen.

Convenience Sample
A convenience sample is selected on the basis of how convenient certain elements of the population are for observation. Convenience may be needed because of time pressures, process accessibility, or other considerations.

Judgment Sample
A judgment sample is obtained based on the discretion of someone familiar with the relevant characteristics of the population. This is usually based on the heuristics of an experienced individual. It may also be based on a group consensus.

Random Sample
A random sample's elements are chosen at random. This is the most important type of sample for statistical analysis. In random sampling, all the items in the population have an equal chance of being selected to be included in the sample.

Because a sample is a collection of observations representing only a portion of the population, the way in which the sample is chosen can significantly affect the adequacy and reliability of the sample. Even after the sample is chosen, the manner in which specific observations are obtained may still affect the validity of the results. The possible bias and errors in the sampling process are hereafter discussed.

Sampling and Nonsampling Errors
A sampling error refers to the difference between a sample mean and the population mean that is due solely to the particular sample elements that are selected for observation. A nonsampling error refers to an error that is due solely to the manner in which the observation is made.

Sampling Bias
A sampling bias refers to the tendency to favor the selection of certain sample elements having specific characteristics. For example, a sampling bias may occur if a sample of the personnel is selected from only the engineering department in a survey addressing the implementation of high-technology processes.

Stratified Sampling

Stratified sampling involves dividing the population into classes, or groups, called strata. The items contained in each stratum are expected to be homogeneous with respect to the characteristics to be studied. A random subsample is taken from each stratum. The subsamples from all the strata are then combined to form the desired overall sample. Stratified sampling is typically used for a heterogeneous population, such as data on employee productivity in an organization.

Through stratification, groups of employees are set up so that the individuals within each stratum are mostly homogenous and the strata are different from one another. As another example, a survey of managers on some important issue of worker involvement may be conducted by forming strata on the basis of the types of processes they are involved with. There may be one stratum for technical processes, one for construction processes, and one for manufacturing processes.

A *proportionate stratified sampling* results if the units in the sample are allocated among the strata in proportion to the relative number of units in each stratum in the population (i.e., an equal sampling ratio is assigned to all strata in a proportionate stratified sampling).

In *disproportionate stratified sampling,* the sampling ratio for each stratum is inversely related to the level of homogeneity of the units in the stratum. The more homogeneous the stratum, the smaller its proportion included in the overall sample. The rationale for using disproportionate stratified sampling is that, when the units in a stratum are more homogeneous, a smaller subsample is needed to ensure good representation. The smaller subsample helps to reduce sampling cost.

Cluster Sampling

Cluster sampling involves the selection of random clusters, or groups, from the population. The desired overall sample is made up of the units in each cluster. Cluster sampling is different from stratified sampling in that differences between clusters are usually small. In addition, the units within each cluster are generally more heterogeneous. Each cluster, also known as a *primary sampling unit,* is expected to be a scaled-down model that gives a good representation of the characteristics of the population.

All the units in each cluster may be included in the overall sample or a subsample of the units in each cluster may be used. If all the units of the selected clusters are included in the overall sample, the procedure is referred to as a *single-stage sampling.*

If a subsample is taken at random from each selected cluster and all units of each subsample are included in the overall sample, then the sampling procedure is called a *two-stage sampling.*

If the sampling procedure involves more than two stages of subsampling, then the procedure is referred to as a *multistage sampling*. Cluster sampling is typically less expensive to implement than stratified sampling. For example, the cost of taking a random sample of 2,000 managers from different industry types may be reduced by first selecting a sample, or cluster, of 25 industries and then selecting 80 managers from each of the 25 industries. This represents a two-stage sampling that will be considerably cheaper than trying to survey 2,000 individuals in several companies in a single-stage procedure.

INTERVIEWING

The most popular and widely used form of review is the interview. In the unstructured method of interview, the auditor sits with the auditee and goes through the operating procedure. The auditee may describe the process verbally only or verbally while performing a task. The auditor records the information and asks spontaneous questions in order to obtain more information concerning the procedures. Some specific methods that are often used during interviews include the following:

1. *Problem discussion:* This explores the kind of data, knowledge, and procedures needed to solve specific problems.
2. *Problem description:* This requires that the auditee describe a problem for each category of answer in the work station.
3. *Problem analysis:* This presents the auditee with a series of realistic problems, and he or she is asked how they are normally handled.
4. *Examination:* This requires that the auditor examine and critique the operating procedures.

An unstructured interview may be first used in the preliminary stage of the knowledge acquisition to obtain a large amount of general information. Later, a structured interview can be used to gain specific information about one particular aspect of the auditee's techniques. It is also useful to record the interview on audio- or videotape. Recording helps to document the interview and also provides a way for the auditor to analyze the auditee's verbal and facial behaviors. However, recording alone is not enough; the auditor must take good notes during interviews. Privacy is very important and interruptions during interview sessions should be kept to a minimum.

Open-ended interviews require either a pilot audit (possibly from unstructured interviews) or an auditor with a significant amount of familiarity with the process to be audited. When a preliminary audit exists, the auditor goes over the contents, making comments on each one. This way additions or deletions may be made very quickly and easily. Tape recording this interview may not be necessary because the auditor can write notes directly on a copy of the pilot audit. The same process can take place when the auditor is very familiar with the process. The interview consists of specific questions directed at certain aspects of the process. The auditee answers the questions and elaborates when necessary. Open-ended questions are posed in a manner that allows for explanations such as Why, When, Who, How, and What?

In general, unstructured interviews have the advantage of generating a large amount of data. This is especially true in the early stages of process evaluation. It is the job of the auditor to keep the auditee from digressing to other unrelated topics and to control the amount and detail of comments. The main disadvantage of interviews is the fact that they are very time-consuming. Unstructured interviews may take weeks to conduct and may be very inefficient because of their informal nature.

Structured interviews may also be very time-consuming because preliminary audits, by nature, may be very long and covering the material may take a long time. This method, however, tends to be more efficient than unstructured interviews.

In terms of the validity of the data, the effectiveness of interviews depends mainly on the skill of the interviewer in asking the right questions and the skill of the auditee in explaining his or her knowledge and techniques. Not all auditors and auditees have this skill. Some of the shortcomings associated with interviews are as follows:

1. There is a tendency to focus on the leading items in a sequence of events when dealing with the entire process sequence.
2. Easily available data sets are often utilized without regard to their relevance.
3. There is a tendency to be conservative in complex decision problems.
4. The manner in which data is presented may affect the ability to perceive the inherent information.
5. Too much unnecessary data complicate the audit and may overwhelm the auditor.
6. People tend to believe a fact if it is thought to be important.
7. There is a tendency to use past successful strategies whether or not they fit new situations.
8. There is a tendency to remember the first and last items in a sequence better than the middle ones.
9. There is a tendency not to explore the subtle aspects of the quality process.

Some suggestions for assisting auditors in obtaining pertinent answers include the following:

- Don't ask all the questions. Some answers are better obtained by observation.
- Trust your instincts. Be prepared to dig deeper into a topic if it appears that you are uncovering a problem area.
- It may be useful to ask the same question of several people to uncover problems with communication.
- Be sure to get the response from the person being interviewed, not the boss, auditee representative, or a staff member.
- Inform people that you will be writing notes in front of them. This is not intended to take direct quotes. Also, note-taking does not necessarily imply a nonconformance but is needed to maintain accuracy or may provide information for the next step in an audit trail.

THE PAPER AUDIT

Questions

Quality Policy

1. Does a written policy on quality exist? Where can one get a copy? How is it disseminated to all employees?
2. What are management objectives for quality? How have they been defined? How are they made available to employees?
3. In what ways does management show its commitment to quality?
4. How has the quality policy been transmitted and explained to all? Are there records, meeting notes, or any documentation of it?

Responsibility and Authority

1. Have personnel whose work affects quality been identified? Who are they (names, location, etc.)?
2. How are responsibilities defined and authority established?
3. Who is the person with authority on product nonconformity?
4. Who is the person with authority to record quality problems?
5. Who is the person with authority to initiate solutions? What are the designated solution channels?
6. How is the effectiveness of these solutions verified?

Verification Resources and Personnel

1. How are in-house verification requirements and activities handled?
2. How are resources for these verification activities provided?
3. Who are the trained personnel assigned to the verification activities?
4. Who performs design reviews and audits? Are they performed by independent personnel?

Management Review

1. How does management review the quality system for effectiveness?
2. Are intervals between reviews specified and observed? What are they? Are the results documented?
3. How are records of these reviews maintained? Where are they stored?
4. Do these reviews include assessment of the results of internal audits?

Quality System

1. Where are the written quality system procedures and instructions? Who develops them?
2. Do the procedures and instructions cover all elements of ISO 9001?
3. How are quality system procedures and instructions being maintained?
4. Has a quality manual been prepared? Who issues it? Is it being used?
5. How are quality plans prepared? What kinds of plans are issued? Are they used?
6. What controls and processes exist to achieve the required quality? Are they in place? Which ones have you seen in action?
7. How is the inspection equipment identified? How much of it has been acquired?
8. How are skills needed to achieve required quality identified? Are they in place?
9. What are the means for updating quality control, inspection, and testing?
10. How is new instrumentation developed to meet changing needs?
11. Is there a means to identify needed improvements?
12. Are standards of acceptability clear?
13. Are design, production process, inspection, and testing compatible?

Contract Review

1. Are there written procedures for contract review? Where are they?

2. How are contracts reviewed to ensure that customer requirements are defined and documented?
3. How are contracts reviewed to ensure that capability to meet customer requirements exists?
4. Are records of contract reviews maintained? Where?
5. Are interface mechanisms between the customer and company clear? How is this process documented? By sign-off?

Design Control

1. What are the written procedures for control and verification of design? Are they being used? Where are they filed?

Design and Development Planning

1. Do project plans exist that identify the responsibility for each design activity?
2. Are these project plans updated as the design evolves? How is an update noted and distributed?

Organizational and Technical Interfaces

1. Are the organizational and technical interfaces identified? How can this be verified?
2. Is project information documented, transmitted, and reviewed regularly? Where can a current example be found?

Design Input

1. Are product/project requirements documented? How? Where are they filed? Extent of distribution?
2. Have the requirements been reviewed for adequacy? What is the review process?
3. Are the requirements clear and consistent? Who makes the determination?

Design Output

1. Is design output documented in terms of requirements? Where can an example be found?
2. Is design output assessed for conformance with design inputs? How is this done and by whom?
3. Does design output contain acceptance criteria? Target values? Tol-

erances? Attributes? Durability? Safety? Reliability? Maintainability? Who determines this and is there an example?
4. Is design output formally assessed for conformance to regulatory requirements? When, how, and by whom?
5. Does design output identify those requirements that are crucial to the safe and proper functioning of the product? Show an example of this type of identification.

Design Changes

1. What are the written procedures used to identify and document all changes?
2. Do these procedures require review and formal approval? How is this done? Who does it? How often?
3. Are design change procedures being used? Are they written? Where are they kept?

Document Control

1. Are there written procedures to control all documents and data related to ISO 9001 requirements? Where can someone get a copy? Are they being used?
2. Have documents falling under these control requirements been identified?
3. Are these controlled documents reviewed and approved prior to use? What is the procedure for this process? Is it documented? Where is it located?
4. Do authorized personnel perform these reviews and approvals? Have authorized personnel been identified?
5. Are documents available to personnel operating the process? On a walk through, has this been observed?
6. Are obsolete documents promptly removed from all points of issue or use? Is this clearly communicated to everyone? How direct is the flow of new documents to the persons needing them?

Document Changes and Modifications

1. Are document changes reviewed and approved by the same groups that performed the original review?
2. Do these groups have access to pertinent background information on which to base their review and approval?
3. Is the nature of the changes identified in the documents?

4. Is there a master list that identifies the current revision of documents? What is it called and where is it located?
5. Is this master list accessible to users so that they do not use nonapplicable documents?
6. Are documents reissued to incorporate accumulated changes?

Purchasing

1. What is the system in place to ensure that incoming material conforms to specified requirements?
2. How are subcontractors selected?
3. Does the type and extent of control over subcontractors vary according to the type of product?
4. Is there a means of ensuring that subcontractor quality system controls are effective? How is this done?
5. Do purchasing documents uniquely describe the product ordered?
6. Are reference documents identified by title and issue?
7. Are requirements for approval or qualification of the product clearly described?
8. Is the quality system standard identified in purchase documents?
9. Are purchasing documents reviewed and approved against documented criteria prior to release?
10. What procedures exist for verification, storage, and maintenance of supplied product?

Product Identification and Traceability

1. What are the written procedures for identifying the product during all stages of production, delivery, and installation? Are they being used?
2. Do individual products have a unique, recorded identification for traceability purposes?

Process Control

1. Are production processes that directly affect quality identified? How?
2. Are there documented work instructions for each of the identified processes? Are they being used? Where are the source documents filed? Who maintains and distributes them?
3. Is suitable equipment used on each of these identified processes? How is this determined?
4. Does a suitable working environment exist for each of these identified processes? How is this determined?

5. Are these processes in compliance with reference standards and codes? Who checks for this?
6. Have important product and process characteristics been defined?
7. Are identified processes and equipment formally approved on the basis of documented criteria? How are these criteria established? Who establishes them?
8. Are there written standards for and/or samples of criteria for workmanship for these identified processes? Where are they kept? How accessible are they?

Inspection and Testing

1. Is incoming material inspected or otherwise verified as conforming to specified requirements prior to use or processing?
2. Is verification of incoming material done in accordance with a quality plan or documented procedures?
3. When incoming material is released for urgent production purposes before verification, is it positively identified and recorded to permit immediate recall and replacement in the event of nonconformance to specified requirements?
4. How is incoming material inspected or verified? Who does it?
5. How are in-process inspection, testing, and identification carried out? Are they done in accordance with quality plan and/or documented procedures? Where are these procedures located?
6. Is conformance with requirements established using in-process monitoring and control methods?
7. Is product held until required inspections and tests have been completed? How is this controlled?
8. When product is not held for required inspections and tests, are positive recall procedures in place?
9. Has it been established that positive recall procedures are reliable?
10. Is nonconforming product identified? How?

Final Inspection and Testing

1. How do final inspection plans/procedures specify that finished product shall have passed earlier inspections and tests?
2. How is final inspection and/or testing carried out? Is it done in accordance with the quality plan or documented procedures?
3. How do final testing and inspection results show evidence of conformance of the finished product to specified requirements?

4. Is finished product held until all required activities have been satisfactorily completed?
5. Is finished product held until final inspection and test data are available and authorized (reviewed and accepted)?

Inspection and Test Records

1. Is there evidence that the finished product has passed the inspections and tests as defined by the acceptance criteria?
2. Are inspection and test records accessible? Where?
3. Have required product measurements been identified? Explain.
4. Has the accuracy of these measurements been defined? Explain.
5. Has inspection, measurement, and test equipment been identified? Does this include on-line devices?
6. Has inspection, measurement, and test equipment been calibrated?
7. Is this equipment adjusted at prescribed intervals, or prior to use, against certified equipment having a known, valid relationship to nationally recognized standards? Explain.
8. Where no calibration standards exist, is the basis for calibration documented? Where?
9. Are there written calibration procedures for each type of equipment? Where?
10. Do calibration procedures include details of equipment type, identification number, location, frequency of checks, check method, acceptance criteria, and action to be taken when results are unsatisfactory?
11. Is there documented evidence that shows that equipment is capable of the accuracy and precision necessary? Where is it kept?
12. Is calibration status shown (e.g., by sticker or record) for inspection, measurement, and test equipment?
13. Do calibration records exist for each piece of equipment? Are they maintained? Where?
14. Is the validity of previous results assessed and documented when equipment is found to be out of calibration? What is the process to handle this? Is it documented? Where?
15. Are calibrations, inspections, measurements, and tests being carried out under suitable environmental conditions?
16. Are the handling, preservation, and storage of this equipment such that accuracy and fitness for use are maintained?
17. Are there safeguards against adjustments that would invalidate calibration settings?

18. Are test hardware (e.g., jigs, patterns) and/or software used for inspection?
19. Are these hardware/software items checked to prove that they are capable of verifying the acceptability of product prior to use?
20. Are these hardware/software items rechecked at prescribed intervals? Have the extent and frequency of these capability checks been defined?
21. Are records of these hardware/software capability checks maintained? Where?

Control of Nonconforming Product

1. Are there written procedures to prevent nonconforming product from inadvertent use or installation? Where are they located? Are they being used?
2. How is nonconforming product identified, evaluated, segregated, and disposed of?
3. Are affected people notified of nonconforming product? How is this done?
4. How is the responsibility and authority for review of nonconforming product defined?
5. Are there written procedures for the review of nonconforming product? Where are they located? Are they being used?
6. How do procedures for review of nonconforming product cover disposition by
    - Rework to meet specified requirements,
    - Acceptance (with or without repair) by concession,
    - Regrading for alternative applications, or
    - Rejection or scrapping?
7. How and when is the proposed use or repair of nonconforming product reported to the customer?
8. Is repaired or reworked product re-inspected?

Corrective Action

1. Are there written procedures for corrective action? Where are they? Are they being used?
2. Do corrective action procedures require identification of the cause of nonconforming product and taking action to prevent (or minimize) recurrence?
3. Are there written procedures for analyzing all processes, work operations, all concessions, quality records, service reports, and customer

complaints to detect and eliminate potential causes of nonconforming product? Where are they? Are they being used?
4. Are corrective actions taken? Are they effective?
5. Are procedures changed as a result of corrective action?

Handling, Storage, Packaging, and Delivery
1. Are there written procedures for handling, storage, packaging, and delivery of product? Where are they? Are they being used?
2. Is product handled in ways that prevent damage and/or deterioration?
3. Are secure storage areas or stock rooms provided that prevent damage or deterioration pending use or delivery?
4. Are methods stipulated for receipt and dispatch to and from storage or stock rooms?
5. Is the condition of stock assessed at periodic intervals to assess deterioration? How often? Is there a checklist?
6. Are the packing, preservation, and marking processes controlled?
7. Is all product identified, preserved, and segregated from time of receipt to end of responsibility?
8. Are there provisions for the protection of product quality after final inspection and test? Does this protection include delivery to destination?

Quality Records
1. Are there written procedures for identifying, collecting, indexing, filing, sorting, maintaining, and disposing of quality records? Where are they kept? Are they being used?
2. Are quality records generated and maintained?
3. Are subcontractor quality records a part of the quality records?
4. Are quality records legible and identifiable to the product involved?
5. Are quality records stored so that they are readily retrievable?
6. Are quality records stored to minimize deterioration or damage and prevent loss?
7. Are retention times for quality records established and recorded?
8. When agreed contractually, are quality records available for evaluation by the purchaser? For how long?

Internal Quality Audits
1. Are internal quality audits performed? When? How frequently?
2. Do internal audits verify whether quality activities comply with planned procedures?

3. Do internal audits evaluate the effectiveness of the quality system?
4. Are audits planned and scheduled?
5. Are audits and follow-up actions carried out in accordance with written procedures?
6. Are the results of audits documented and brought to the attention of personnel having responsibility in the area studied?
7. Is corrective action taken on any deficiencies found by the audit?

Training

1. Are there written procedures for identifying training needs for activities affecting quality? Have activities affecting quality been defined?
2. Is identified training provided for all personnel performing activities affecting quality?
3. Are personnel performing specific assigned tasks qualified on the basis of appropriate education, training, and/or experience?
4. Are records of training maintained?

Servicing

1. Are there written procedures for performing and verifying specified service requirements?

Statistical Techniques

1. Have statistical techniques been identified? Are they being used properly? Are outputs interpreted correctly?

IS THIS THE END OF THE JOURNEY?

No, the journey of quality improvement has no end. There is always room for more improvement. This book has presented concepts and tools that can enable an organization to embark on the journey to improving operating procedures and achieving ISO 9000 certification. For cases where formal certification is not being sought, the lessons and ideas contained in this book are still useful for general improvement. The topics covered are universally applicable. It is one thing to dream, plan, and hope, but it is a different thing to actually implement. The techniques of project management, as recommended in this book, offer real opportunities for quality improvement.

Higher Quality to You!

APPENDIX A
SOURCES OF ISO 9000 INFORMATION AND RESOURCES

ABCS Consultants
1302 Vine Street
Norman, OK 73072
Tel: 405-321-3426
   Custom in-house project management training for ISO 9000

Applied Computer Services, Inc.
P.O. Box 2193
Norwalk, CT 06852
Tel: 203-849-9557
FAX: 203-849-9890
   Software tools for ISO 9000

ASQC (American Society for Quality Control)
Tel: 414-272-8575
FAX: 414-272-1734
   Official copies of the ISO 9000 standard

Business Challenge, Inc.
385 South End Avenue, Bldg. 500, Suite 3G
New York, NY 10280
Tel: 212-488-7038
FAX: 212-488-9479
   Company assessment system for ISO 9000

CEEM
10521 Braddock Road
Fairfax, VA 22032-2236
Tel: 800-745-5565
FAX: 703-250-5313
  ISO 9000 registered companies data

Center for Organizational Improvement
900 N. Portland Avenue
Oklahoma City, OK 73107-6195
Tel: 405-945-3278
FAX: 405-945-3397
  ISO 9000 training, general related information

Center for Professional Advancement
P.O. Box 1052
East Brunswick, NJ 08816-1052
Tel: 908-613-4535
FAX: 908-238-9113
  ISO 9000 training

Efe Quality House
3970 Chainbridge Road
Fairfax, Virginia 22030
Tel: 703-359-5969, 703-691-1697, 703-691-9850
FAX: 703-359-5971, 703-691-9399
  ISO 9000 training, general related information

Foxboro's ISO 9000 Register
Tel: 713-579-9003
FAX: 713-579-1360
  General information on ISO 9000

Institute of Industrial Engineers
25 Technology Park/Atlanta
Norcross, GA 30092
Tel: 404-449-0460
FAX: 404-263-8532
  ISO 9000 reference materials

ISO 9000 Training Center
Moore-Norman Vo-Tech
4701 12th Avenue, NW
Norman, OK 73069
Tel: 405-364-5763 (ext. 288)
  ISO training workshops

Motion Knowledge Systems Ltd
2860 Porter Street
Soquel, CA 95073
Tel: 408-479-4074
FAX: 408-479-1007
E-mail: mksltd@aol.com
   ISO 9000 software tools, self-assessment tools

National ISO 9000 Support Group
9964 Cherry Valley, Bldg. #2
Caledonia, MI 49316
Tel: 616-891-9114
   ISO 9000 support, computer bulletin board

Quality Alert Institute
ISO 9000 Group
P.O. Box 43155
Upper Montclair, NJ 07043-7155
Tel: 800-221-2114
   Custom in-house ISO 9000 programs

Quality Systems Update
Tel: 703-250-5900
FAX: 703-250-5313
   ISO 9000 news and directory of registered companies

Qualtec Quality Services, Inc.
Tel: 407-775-8300
FAX: 407-775-8301
   Quality improvement programs and ISO 9000 training

Society of Manufacturing Engineers
One SME Drive
P.O. Box 6028
Dearborn, MI 48121
Tel: 800-733-4763, 313-271-1500
FAX: 313-271-2861
   ISO 9000 training materials

APPENDIX B
GLOSSARY OF INTERNATIONAL BUSINESS TERMS

Abrogate. To abolish or declare void by formal authoritative measures.

Acceleration clause. A condition in a loan or mortgage that makes the whole of the outstanding balance due in the event of failure by the debtor to maintain regular payments.

Acceptance supra protest. When a bill of exchange is protested and then accepted by another party to save the name of the drawer, this is called acceptance supra protest, or acceptance for honor.

Accommodation note. A document signed by an individual who is prepared to act as a guarantor on behalf of a person whose credit is doubtful.

Acknowledgment. A formal notice to the sender of goods or money acknowledging that the shipment has been received.

Advice note. Usually notes sent by railways for freight companies to inform firms that their consignments have arrived and are awaiting collection.

Airway bill. An air transport term for the document made out on behalf of the shipper as evidence of the contract of carriage. It is also called an air consignment note.

Annual return. The annual return must show names of persons who have ceased to be owners since the last return, a statement showing the sales and profits of the business for the year, and a statement of the assets and capital of the business at the end of the year. It is open for inspection by any person.

Appropriation account. The account of a business that shows how the profits are to be allocated.

Articles of association. The document containing the conditions of regulations for the conduct of the internal affairs of a company.

Assets. Items of value whose value arises not from any intrinsic worth but from their ability to earn revenue.

Assets (current). Those assets that form part of a company's trading cycle (e.g., stocks, raw materials, etc.).

Assets (fixed). Those assets that enable a company to carry out its operations are called fixed assets. They form the basic production facilities of a company (e.g., land, factories, offices, machinery, etc.). Most are tangible, but some are intangible (e.g., patents).

Average costs. The total production expenses (including minimum profit) divided by the number of commodities produced.

Balance Sheet. A statement produced periodically, normally at the end of the financial year, showing an organization's assets and liabilities, expressed either as totals or as balances if a two-way flow has occurred.

Balloon note. A promissory note that necessitates small repayments during the early period of a loan and larger repayments toward the latter period.

B/D. Bank Draft.

Bill of exchange. Defined officially as "an unconditional order in writing addressed by one person to another, signed by the person giving it, requiring the person to whom it is addressed to pay on demand, or at a fixed or determinable future time, a certain sum in money to, or to the order of a specified person, or to the bearer" (Bills of Exchange Act 1882), a bill of exchange is essentially a post-dated check, drawn by a supplier in his own favor to be signed by a company.

Bill of lading. A bill of lading serves three purposes: (1) When signed by the ship owners, it acknowledges the safe receipt of the goods on board ship. (2) It constitutes the contract between ship owner and shipper and states to what extent the ship owners can be held responsible for safe carriage of the goods. (3) The bill is also the document of title and the holder is entitled to demand delivery of the merchandise on producing it.

B/L. Bill of Lading. Receipt given on behalf of ship owner for goods shipped or received for shipment.

Blanket insurance. Insurance cover that relates to a range of goods or property that may vary at different periods of time.

B/P. Bill Payable.

Bridging loan. A short-term advance by a bank to a customer pending the receipt of funds from another source—e.g., when a person buying a new house has to wait for the old house to be sold.

Buffer stock. A quantity of raw material retained in store to safeguard against unexpected shortages because of delivery delays or sudden upsurge in demand.

Ceiling prices. The maximum prices that are permitted when price-control legislation is in effect.

Certificate. A written declaration of the truth of a specific matter.

C.I.F.: Cost Insurance Freight. The seller undertakes to arrange to supply goods, pay the freight charges, and insure them until they reach the destination, for a quoted price.

COB. Close of Business.

Condition. A term in a contract; failure to fulfill allows the other party to repudiate. It must be a statement of fact.

Consignee. The party to whom goods are being sent.

Consignor. The sender of a consignment of goods.

Continuation clause. A marine insurance clause covering situations where a ship is still at sea when its insurance runs out. The insurer agrees to cover risks at a pro rata rate of premium.

Contract. A legally enforceable agreement between two or more persons. It normally takes the form of one person's promise to do something in consideration of the other's agreeing to do or suffer to do something in return.

Copyright. The exclusive right that is legally granted to the author to make and control copies of a book or other artistic work.

Debentures. These represent loans to a company as opposed to shares in it, as a form of investment. The holders have no say in the company policy or in the election of directors; they are, however, entitled to interest on their loans before profits are distributed. If the company fails to make these payments, debenture holders would be entitled to sell up the business in order to obtain their due interest and loan repayment.

Debit Notes. Notes informing a customer that his or her account in the seller's ledger has been charged with a stated sum, and the reason for this. Debit notes are used to adjust undercharges, as credit notes are used to adjust overcharges.

Depreciation. A term for the amount by which the usefulness of a fixed asset has diminished.

Dividend. A proportion of company profits paid to shareholders at regular intervals.

Dividend warrant. A check in payment of a dividend, issued by a company to shareholder.

Down market. A term used to describe goods that are aimed at the lower-quality end of the market.

Dry goods. A collective term for clothing, hardware, and related merchandise as distinct from grocery items.

Duty. A tax levied on goods as a means of producing revenue for the country.

EEC. European Economic Community.

EFTA. European Free Trade Association.

Ejectment. A legal action for the recovery of the possession of property.

Elastic demand. Demand is elastic when a small change in price results in a large change in demand.

Endorsement. A signature written on the back of a document or a bill of exchange so that it is transferred to a third party.

Endowment fund. A fund from which the interest may be spent but the capital sum must be maintained.

Escalator clause. A condition in a contract that allows for increases in price in certain specific circumstances.

Estimate. An approximate assessment of the price of goods or services given in advance by those who wish to undertake their supply.

Executed contract. A contract where one party has performed all that is required to fulfill the obligations incurred.

Exemption clause. A clause, often in a sale-of-goods contract, excluding the supplier from any claims as to the condition of the goods.

Extended coverage. An additional clause in an insurance policy that provides coverage against risks in addition to those stated in the basic policy.

F.A.S. Free Alongside Ship. Indicates that the quoted price includes all costs up to bringing the goods alongside the vessel. The importer must then pay for goods to be taken on board and all freight and insurance charges.

Fixed charge. Unavoidable overheads such as interest payments, depreciation, and rent, which cannot be related to the level of business activity of an organization.

Fixed costs. Overhead costs that do not vary with the volume of production (e.g., rent).

F.O.B. Free on Board. Indicates that the cost quoted includes all expenses up to and including the loading of the goods on the vessel at a specified port.

Free from particular average. A marine insurance term meaning that the insurers are not responsible for anything other than total loss and general average loss.

Free of all average. A term for policies where the insurer takes no responsibility for general average loss and will pay only on total loss.

Free trade. International trade that is unrestricted by import and export quotas, tariffs and other controls that impede the free movement of goods between countries.

Freight absorption. The practice of a seller of goods not charging the purchaser for costs incurred in the transport of goods.

Fundamental disequilibrium. The constant discrepancy between the official exchange rate of the currency of a country and purchasing power of that currency.

G.B.O. Goods in Bad Order.

GATT. General Agreement on Tariffs and Trade. Signed in 1947 by a group of countries and directed toward the reduction of trade barriers.

General average. A marine insurance term for a loss that has been deliberately incurred in order to avoid a danger—e.g., when cargo is thrown overboard to prevent a ship from sinking. In such instances the loss is apportioned between the ship owners and cargo owners.

General tariff. A standard rate of duty applied to an unparted commodity irrespective of the country of its origin.

Going long. The practice in the commodity markets of buying first and then selling, in the hope of a rising price trend.

Going short. The commodity market term for selling before purchasing, with a view to profiting by a fall in prices.

Gross weight. The combined weight of goods and packaging.

Groundage. A charge levied by some port authorities for permitting vessels to anchor there.

Guarantee. An agreement to be responsible for the debt, default, or miscarriage of another.

Guarantor. A person who makes or gives a guarantee.

Harbor dues. A payment made for use of port facilities.

Hard sell. A marketing term denoting a vigorous sales campaign, a procedure that is adopted when selling is difficult.

Hedging. Protecting an investment against loss caused by fluctuations in the market. This is usually achieved by offsetting the cost of a present purchase against a purchase of the same commodity or security in the future.

Holder. The person who has possession of a negotiable instrument.

Holding company. A company formed for the express purpose of exercising financial control over a number of operating companies by buying up all or the majority of their shares.

Impost. Another name for a tax, especially one levied on imports.

Indemnity. An agreement whereby one person agrees to make good any loss suffered by a party to a contract to which he himself is a stranger. The most common contracts of indemnity are insurance contracts.

Indenture. A form of agreement which is drawn up for the sale of bonds or debentures.

Instrumental capital. All capital equipment that is used in the production of goods.

Interest warrant. A document sent to shareholders that entitles them to the payment of interest on stock.

Invoice. A document issued by a vendor of goods stating the nature of those goods, the name of the debtor, and the sum due.

Jettison. A marine insurance term meaning to throw goods overboard for a good and sufficient reason—e.g., in order to lighten ship in time of danger.

Job costing. The allocation of all costs to a specific contract for a particular customer.

Judgment note. A legal document acknowledging the existence of a debt.

Keelage. The payment made by a shipowner when a ship enters and remains in port.

Kickback. Illegal payment made to obtain a contract or favor.

Kiting. Taking advantage of the check clearing system by drawing checks on deposits that are already committed, assuming that the delay in clearing the check will allow time to replenish the account.

Landed terms. A quotation for the sale and carriage of goods by sea that includes all costs up until the goods are actually landed.

Lay days. A specified number of days allocated for the unloading and loading of a ship, commencing with arrival in port.

Lease. A contract specifying the terms, of time and cost, of possessing and using a property.

Lessee. The person to whom a lease is granted.

Letter of credit. A document issued by a bank that entitles the bearer to draw money on the bank's account. Often used for overseas transactions.

Leverage. The ratio between the debts of the business and the owner's capital.

Limited liability. A limited liability company is, at law, a separate entity quite apart from the people who own it, and only the company can be sued. It follows that the liability of the shareholders of such a company is limited to the amount of capital they have invested in it.

Limited partner. A limited partner cannot take part in the running of the business; each partner's liability is limited to the amount he or she invested.

Liquidated damages. The amount calculated as recompense if a contract is not fulfilled.

Liquidation. This is effectively the bankruptcy of a company. There are two forms of liquidation: voluntary and compulsory. When a company is wound up, assets are realized and debts paid as far as possible. If any monies remain, these are distributed to the owners of the company according to their respective rights.

Marginal costing. Calculating the production costs excluding all fixed costs and overheads, which do not alter with the volume of production.

Marked check. A check that is guaranteed by the bank on which it is drawn.

Merger. The combination of two or more companies to form one new company.

Monopoly. Any firm that produces such a high proportion of the total output of a commodity that it can influence price by regulating supply is said to have a monopoly of that commodity.

Monopsony. A market situation where there is only one major purchaser of goods or services.

Mortgage debenture. The commonest fixed interest security is a mortgage debenture. This is usually secured on the most saleable assets of the company—e.g., head office, land, etc.

NAFTA. North American Free Trade Agreement. Trade agreement between the United States, Mexico, and Canada.

Not negotiable. The cancellation of the free transferability of a negotiable instrument usually by writing the phrase on the face of the document.

Notice. A period of time granted by law to an employee who has been made redundant or dismissed during which he must continue to be employed at his original job or alternatively given wages in lieu of notice.

Objective value. In economic terms this phrase implies the value in exchange or the market value of a product. A product can have value to the owner without being of value to anyone else.

Oligopsony. A market situation where there are only a few purchasers of certain goods or services so that each purchaser is in a position to influence the price paid.

Open check. A check that is uncrossed and can therefore be cashed over the counter without being first being paid into an account.

ORD. Owner's Risk of Damage.

Ordinary shares. The customary form of equity investment in a company with a par value. Ordinary shareholders have no guaranteed claim on profits, but they receive dividends from profits on the recommendation of directors and may vote at the election of those directors.

Original bill. A bill of exchange sold before it has been endorsed.

Pari passu. In equal proportions.

Partnership. A relationship that exists between two or more people to carry on a business in common with a view to profit.

Patent. A license taken out by an inventor to protect his or her invention from unauthorized use by other persons.

P.O.D. Pay on Delivery.

Power of attorney. A form of instrument authorizing the holder to act on behalf of another—e.g., to sign deeds. This is often used when the principal is in a different country.

Preference shares. As opposed to ordinary shares, these carry a fixed rate of dividend payable as a first charge out of profits.

Primage. A freight charge paid in order to ensure that care is taken when loading or unloading goods.

Prime costs. Those that vary with output, as opposed to fixed costs, which are standing charges that do not change with increase or decrease of output.

Pro forma invoice. A preliminary invoice stating the value of goods and notifying the recipient that they have been dispatched. It is not a demand for money and might for instance accompany goods sent off on approval.

Pro rata. In proportion to a total sum.

Prohibited risk. Not insurable because the risk is too great for the insurance company to accept.

Promissory note. A document containing a promise to pay a certain sum of money at a fixed time. It can become a negotiable instrument.

Proof of loss. The written document required by an insurance company as proof that a loss has taken place.

Proviso. A condition in a document or contract.

Qualification shares. Shares that a director of a company is obliged to own if he or she is to be recognized as a director.

Quick ratio. The relationship between current assets and current liabilities. This should indicate if a business can pay its liabilities quickly in cash.

Quorum. The number of persons who must be present at a meeting in order that it may officially take place.

Quota. A limit set on the entry of goods or persons into a country.

Quotation. A definition in writing of the price, terms, and conditions attaching to a potential contract.

Rally. An improvement in the market price of shares or commodities after a decline.

Rate of exchange. The rate at which one currency will exchange for another.

Ratification. The formal confirmation of an agreement applied especially in the law relating to agents.

Realization. Conversion of assets into cash.

Receiving order. The first stage in bankruptcy proceedings. It is a court order that brings the property of a debtor under the protection of the court until the bankruptcy proceedings have been concluded.

Redraft. A new bill of exchange that is drawn up when a bill is dishonored. It incorporates the additional charges involved in the dishonor.

Remittance. Money in any form sent from one person to another.

Repudiation. Formal refusal to pay a debt by the debtor.

Rights issue. With a rights issue of ordinary shares, new shares are offered to the holders of existing ordinary shares in proportion to their existing holdings. The right is a preemptive choice of purchase of the new shares.

Risk. A risk situation exists if the outcomes of a decision can be estimated, and each outcome can be estimated to have a probability of occurring. This is to be distinguished from an uncertainty situation when the outcomes can be estimated but the probability of each outcome occurring is unknown, and a situation of complete uncertainty, when neither outcomes nor probabilities can be estimated.

Royalty. A regular payment made for the use of land for mining, or the use of a patent or trademark, or the right to publish and sell copyright material.

Scab. A worker who refuses to cooperate with the unions in a strike.

Scrip issue. Free shares issued to shareholders in order to reduce reserves.

Search warrant. A legal document that authorizes a person to enter private property or open packages in the pursuit of stolen goods.

Security. Something given or guaranteed by a borrower as safeguard for a loan. The term is not applied to shares but to debentures and similar loan stock and to negotiable instruments.

Severable contracts. This phrase often occurs in contracts for the sale of goods, particularly when the goods are to be delivered in installments. It is possible, but not necessary, to treat separate installments as separate contracts.

Sight bill. A bill that becomes payable as soon as it is presented.

Sleeper. A security or business that has not been active, but is believed to be potentially valuable.

S/N. Shipping Note.

Spot price. Current market price.

Stale check. One that has been unpaid or not presented for a considerable time.

Stamp duty. A tax applied to legal documents.

Subpoena. A legal writ that compels a person to appear in court.

Sunk cost. A fixed overhead of a business such as buildings, roads, special machines.

Surtax. Extra tax paid usually on incomes over a certain level.

Tare. The weight of the package or container in which goods are held.

Time bill. Bill of exchange with a fixed date of payment.

Title deeds. The documents that act as proof of ownership of property.

Toll. A charge made for the use of a dock, motorway, bridge, tunnel, etc.

Trademark. An identifying mark than distinguishes the product of one firm from another in the same field.

Turnover. An accountancy term for gross takings or total sales before any deductions are made.

Underwriting. The commitment on the part of a collection of investing institutions or issuing houses to take up shares at the issue prices if other investors fail to do so, in exchange for a fixed fee of between 1 percent and 3 percent of the issue.

Undue influence. A form of moral pressure. A contract will not be enforced by a court if it can be shown that the defendant was in a position that prevented the forming of a free and unfettered judgment. Undue influence will not be presumed unless certain relations (e.g., parental or confidential) exist between the parties.

Unrealized profits. Profits that exist but have not yet been turned into cash.

Unsecured loan. This is a type of fixed interest security not secured to any assets by the deed covering the loan, but there is still the right to realize the assets of the company should it not comply with the conditions laid down in the deed.

Upset price. The minimum price that a seller will accept for a piece of property.

Venture capital. High-risk investment—e.g., shares bought in a new business.

Vertical integration. The linking up of a number of business concerns each operating at a different stage in the same industry. This is opposed to horizontal integration, the linking up of a number of firms at the same stage.

Vested interest. Interest now in being, as opposed to interest anticipated. Property is said to vest when the absolute owner is finally established, and his or her interest is in no way capable of being terminated by anyone else.

Visa. An endorsement on a passport that permits a person to travel in a particular country where entry restrictions are in force.

Visible items of trade. Physical imports and exports excluding the invisibles such as insurance, freight, and bank charges, etc.

Voidable contract. A contract that can be declared void on the basis of insanity, minority, or incompetency of one of the parties to it.

Warranty. A statement of fact in a contract, either express or implied. If it is unfulfilled the injured party cannot repudiate the contract but may be able to claim damages. The difference between a warranty and a condition is that a condition is fundamental to a contract, whereas a warranty is not.

Wasting asset. An accounting term for assets that are used up gradually in producing goods. It is sometimes applied to fixed assets generally, but is perhaps better applied to assets that are exhausted after a certain period of time—e.g., quarries, mines, etc.

Without recourse. A phrase used to protect the seller of a bill of exchange in the event that the bill is unpaid when due.

Work in progress. Represents the value of work commenced but not yet completed (e.g., in a manufacturing business, this will be in partly finished goods; in a contracting business, in the form of uncompleted contracts).

Working assets. Those assets that can readily be turned into cash, such as raw materials, semi-finished goods, debts, etc.

Working capital. A term for that part of a company's capital that is continually circulating. The figure is calculated by deducting current liabilities from current assets.

Writ. A court order.

Write-down. An accountancy expression describing the process of reducing the book value of an asset.

Write-off. An accountancy term denoting the debt that cannot be collected and is therefore written off against profits. It is also more generally applied to any property that has been damaged and is unable to be repaired.

APPENDIX C
GLOSSARY OF QUALITY-RELATED TERMS

Accuracy. A description of how closely the measurements of an instrument match the true values of the parameters being measured.

Analysis of variance (ANOVA). A statistical technique by which the total variation of a set of data is subdivided into component parts, each of which is associated with a specific source of variation. The purpose of ANOVA is to test some hypothesis on the parameters of the model or to estimate variance components.

Appraisal. A form of the quality system audit, normally conducted to examine the total quality program effectiveness and implementation. An appraisal is usually conducted by a third party and reported to the highest level of management.

AQC. Achieving Quick Changeover.

Assessment. An estimate or determination of the significance, importance, or value of something. It is another term for quality audit. It is sometimes used to indicate a less formal means of measuring and reporting than the normal audit. An assessment is usually limited in scope. Assessments may be performed during the contracting phase or the contract execution phase.

Assignable cause. A detectable factor that contributes to variations in a process.

Attribute. A characteristic that is classified in terms of whether or not it meets a given requirement (e.g., go, no go).

Audit. A planned, independent, and documented assessment to determine whether agreed-upon requirements are met. Audit and survey are sometimes used interchangeably. Audit implies the existence of some agreed-upon criteria against which the plans and execution can be checked. Survey implies the inclusion of matters not covered by agreed-upon criteria.

Audit program. The organizational structure, commitment, and documented methods used to plan and perform audits.

Auditee. An organization to be audited.

Auditing organization. A unit that carries out audits through its employees. This unit may be a department of the auditee, a client, or an independent third party.

Auditor. The individual who carries out the audit.

AZD. Achieving Zero Defects.

Batch (or lot). A collection of units of a product manufactured by one supplier under manufacturing conditions that are presumed to be uniform and consistent.

c chart (count chart). A control chart for evaluating the stability of a process in terms of the number (count) of occurrences of a particular event in a sample.

Calibration (based on analysis). The process of establishing the accuracy of a measuring instrument based on an analysis within the operating environment.

Calibration (based on comparison). The process of comparing the accuracy of one measuring instrument against another measuring instrument (standard or gauge) for the purpose of adjusting the accuracy of the subject instrument.

Central line. A line on a control chart representing the long-term central tendency of a parameter being studied.

Certainty. The degree of assurance with which some quantity may be estimated.

Certification. The procedure and action, by a duly authorized body, of determining, verifying, and attesting in writing to the qualifications of personnel, processes, procedures, or items in accordance with applicable requirements.

Certification (of supplier). The process by which products are obtained from a supplier and tested for the purpose of documenting the units from the supplier as qualified or certified products. It is alternately referred to as *qualification*.

CFM. Continuous Flow Manufacturing.

Chance causes (random causes). Random and undetectable factors that contribute to variations in a process.

Change in mean. A measure of the deviation in long-term central tendency of a process from a reference point.

Characteristic. A distinguishing property of the units of a product. A property that helps identify or to differentiate between entities and that can be described or measured to determine conformance or nonconformance to requirements.

Class boundaries. The end points of the possible values of a parameter that is classified into classes.

Class interval. The width of each bar in a frequency histogram.

Class limits. The upper and lower bounds of the class into which items are classified.

Client. The person or organization requesting the audit. Depending on the circumstances, the client may be the auditing organization, the auditee, or a third party.

CMI. Continuous Measurable Improvement.

Compliance. An affirmative indication or judgment that a product or service has met the requirements of the relevant specifications, contract, or regulation; also the state of meeting the requirements.

Component. A single identifiable part of an assembled product.

Confidence interval. A statistical interval that has a certain level of chance of containing the representative value of a parameter.

Confidence level. The probability (or level of confidence) associated with a confidence interval.

Confidence limits. The end points of a confidence interval.

Conformance. The affirmation that a product meets certain requirements. An affirmative indication or judgment that a product or service has met the requirements of the relevant specifications, contract, or regulation; also the state of meeting the requirements.

Conformity. The state of meeting specified requirements by a product.

Consumer's risk. The probability of accepting a bad batch from a producer's production lot.

Contingency table. A tabulation of rows and columns to display relationships between various factors.

Contractor. Any organization under contract to furnish items or service; a vendor, supplier, subcontractor, or fabricator where appropriate.

Control. The corrective actions by which some desired result is ensured.

Control chart. A chart on which limits are drawn and on which are plotted values of a statistic obtained from sequential samples of a product.

Control chart factor. A factor, usually varying with sample size, to convert specified statistics or parameters into a central measure or control limits relevant to the control chart.

Control charts for individual observations. A control chart in which individual observations are evaluated in order to assess the stability of a process.

Control limits. Limits on a control chart that are used as criteria for indicating the state of statistical control of a process.

Control system. The system of controls by which control of some desired result is achieved.

Convention. A customary practice, rule, or method.

Correction factor. An adjustment associated with the measurement of parameter values from an origin other than their mean.

Corrective action. Steps taken to correct adverse conditions affecting a process. Action taken to eliminate the root cause(s) and symptom(s) of an existing undesirable deviation or nonconformity to prevent recurrence.

Corrective action request. A formal document noting audit findings and requesting resolution.

CPI. Continuous Process Improvement.

Cumulative count control chart (CCC chart). A graphical method used for monitoring a process by plotting the cumulative count of conforming units instead of nonconforming units.

Cumulative frequency. The total frequency of parameter values that are located below or at a class boundary.

Customer feedback and corrective action. A program of feedback and corrective action based on customer-supplied data.

Cycle. A single performance of a complete set of operating conditions.

Cycle time. The elapsed time between the commencement and completion of a cycle.

Data. Statistics obtained from measurements and observations in a process.

Defect. A deficiency associated with the failure to meet requirements imposed on a product with respect to a single quality characteristic.

Defective. The state of exceeding the maximum number of allowable defects in a product.

Degrees of freedom. An integer number representing the result of subtracting the number of independent parameters computed in a statistical test from the sample size. It is used for entering statistical tables.

Dependant variable. A characteristic or variable whose value is directly influenced by the values of some other variables.

Design. The representation of ideas, concepts, or plans in a written form.

Destructive testing. Diagnostic tests that stress the characteristics of a product or process to the point of destruction.

Deviation. Any departure from the specified range of a characteristic. A nonconformance or departure of a characteristic from specified product, process, or systems requirements.

DFA. Design for Assembly.

DFM. Design for Manufacture.

Dispersion. The degree of spread present in a set of observations.

Distribution curve. The line enveloping a frequency distribution.

DOE. Design of Experiments.

Drift. The change over time in a characteristic of a product.

Effect of a factor. The change in a response variable based on a change in the level of a factor.

Element. Any component of an audit program. Also refers to any one of the 20 paragraphs of ISO 9001 or 18 paragraphs of ISO 9002.

Estimated process average. Expected average yield of a process based on a sample from the process.

Evolutionary operation (EVOP). An experimental procedure for collecting information to improve a process without disturbing production.

Exit meeting. The meeting at the end of the audit between the auditors and the representative auditees, at which time a rough draft of audit findings and observations is presented.

Experimental design. The planning of experiments to collect statistically valid data and generate statistically valid analysis by varying factor levels under controlled conditions.

Factor. A variable that affects the response or output of a process.

Factor level. A specific setting of the value of a factor in an experimental study.

Factorial experiment. An experiment designed to determine the presence or absence of interactions and assess the effects of one or more factors when each factor is studied at a minimum of two levels. In a full factorial experiment all combinations of all factor levels are investigated.

Feedback loop. A sequence of forward and backward communication on process performance used as an input to perpetrate or improve process stability.

5S's. Seiri, Seiton, Seiso, Seiketsu, Shitsuke (organization, neatness, cleaning, standardization, discipline).

Follow-up audit. An audit that verifies that some corrective action has been accomplished as scheduled and determines that the action was effective in preventing or minimizing recurrence.

Fraction defective. The total number of defective units divided by the total number of items.

Frequency. The number occurrences of a given type of event or the number of members of a population falling into a specified class.

Frequency distribution. A measure of the frequency of occurrence of the respective values of a process parameter.

Frequency polygon. The line enveloping a frequency histogram and passing through the midpoint of each class interval.

Gantt chart. A chart that shows the task schedule in a process. It may also display the status of various tasks, personnel assignments, and costs.

Gauge. A measuring device used to measure the physical dimensions of a product.

Guidelines. Documented instructions that are considered good practice but that are not mandatory.

Histogram. A graphical representation of the class distribution of the values of a process parameter.

Homogeneous. The state of having characteristics that are uniformly distributed throughout a process or a sample from the process.

Independence. Freedom from bias and external influences.

Independent variable. A variable characteristic of an item whose value is independent of the values of other characteristics of the item.

Inherent process variability. The variability that is inherent in a process when operating in a state of statistical control.

Inner noise. Internal factors, such as wear and tear, that influence the functionality and variability of a product.

Inspection. The process of checking the characteristics of a product for conformity. Activities, such as measuring, examining, testing, that gauge one or more characteristics of a product or service and comparison of these with specified requirements to determine conformity.

Inspection by attributes. An inspection in which the item is classified as either defective or nondefective (conforming or nonconforming) or the number of defects with respect to a given requirement.

Inspection by variables. Inspection in which certain quality characteristics of the item are evaluated with respect to a scale of measurements and expressed as divisions along the scale.

Inspection system. The established program by which the inspection of a product is carried out. The inspection system encompasses personnel, equipment, and procedures.

Interaction effect. The effect produced by the interaction between two or more factors that affect the characteristics of a product.

JIT. Just in Time.

Kurtosis coefficient. The degree to which a distribution is flattened or peaked.

Lead auditor. The individual who supervises auditors during an audit as a team leader.

Loss function. A continuous cost function that measures the cost impact of the variability of a product.

Lower control limit (LCL). The lower of the two control limits governing the conformity of an item on a control chart.

Main effect. The effect of a factor acting on its own from one level to another to affect the outcome in a factorial experiment.

Measurement error. Error in a recorded observation because of measurement inaccuracies.

Measurement standard. A standard against which the output of a measuring instrument is compared.

Measuring system. The collection of physical elements used to obtain a measurement of the characteristics of an object.

Median. The value within a distribution above and below which an equal number of parameter values fall.

Metrology. The filed or function associated with the processes of measurement.

MIPS. Minimum Inventory Production System.

Mode. The value within a statistical distribution that has the greatest frequency of occurrence.

Modified control chart. A control chart with modified limits based on subgroup average.

Moving range control chart. A control chart in which the range of the latest n observations is used for evaluating the stability of the variability in a process.

Multivariate control chart. A control chart for evaluating the stability of a process in terms of the levels of two or more process parameters.

Nested experiment. An experiment in which the level of one factor is chosen within the levels of another factor.

Noise factor. An extraneous factor that disturbs the function of a product.

Nominal value. The desired value of a process parameter from which variations are measured in terms of tolerance limits.

Nonconforming unit. A unit of product or service containing at least one nonconformity.

Nonconformity. A departure of a quality characteristic from its intended level in such a way that the departure adversely affects the overall quality of a product.

Normal cost. The minimum activity cost associated with the performance of an activity at the minimum feasible resource level.

Normal time. The minimum time in which an activity can be performed at the normal cost.

***np* chart.** A control chart for evaluating the stability of a process in terms of the total number of units in a sample that meet a certain classification.

Objective evidence. Verifiable qualitative or quantitative observations, information, records, or statements of fact pertaining to the quality of an item or service or to the existence and implementation of a quality system element.

Observation. An item of objective evidence found during an audit.

100 percent inspection. Inspection of all the units in a lot or batch. Also called *screening inspection.*

Optimistic activity time. The expected activity duration that would occur if every aspect of the performance of the activity goes as well as it possibly could.

Orthogonal matrix (array). A fractional matrix that assures a balanced and fair comparison of the levels of a factor in an experimental study such that the columns in the matrix can be evaluated independent of one another.

Out of control. The condition describing a process from which all special causes of variation have not been eliminated. This condition is evident on a control chart by nonrandom patterns within the control limits.

Outer noise. Ambient noise factors (e.g., temperature, humidity, etc.) that affect variation within a process.

***p* (proportion).** A ratio indicating the number of units of a product that meet a certain classification in a sample of the product.

***p* chart.** A control chart for evaluating the stability of the process in terms of the proportion of a sample meeting a certain classification.

Pareto chart. A graphical tool for percentage classification of the potential problem areas in a process according to their contribution to a specified criterion of measure such as cost or number of defects.

Percent defective. The fraction defective multiplied by 100.

Percent nonconforming. The fraction nonconforming multiplied by 100.

Pilot line. A production line set up to collect information on a proposed production system.

Pilot lot. A small batch of a product sampled from a pilot line for the purpose of studying the characteristics of a product or process.

Pooling. The combination of the sum of squares and the degrees of freedom of factors in an analysis of variance to obtain a better estimate of experimental error.

Population. The universal collection of similar items from which samples are drawn for measurement and statistical analysis.

Pre-audit meeting. The introductory meeting between the auditors and the representative auditees, at which time the overview of the planned audit is presented.

Pre-award survey. An activity conducted prior to a contract award and used to evaluate the overall quality capability of a prospective supplier or contractor.

Procedure. A document that specifies the way to perform an activity.

Process. One event or a collection of events within which people, tools, and materials interact to perform operations that cause one or more characteristics of a raw material to be altered or generated.

Process average. The average value of a process in terms of the percentage or proportion of variant units.

Process capability. A standardized evaluation of the inherent ability of a process to perform specified operations after significant causes of variation have been eliminated. Process capability usually is set equal to six standard deviations of the variability.

Process capability study. A controlled collection of statistics from a process for the purpose of statistically determining the capability of the process to produce acceptable products under specified conditions.

Process quality. A statistical measure of the quality of product from a given process.

Process quality audit. A quantitative assessment of conformance to required product characteristics.

Process spread. The total variability that exists in items produced by the process.

Process tolerance. The range over which the values of a characteristic of the product from a process are allowed to vary. Process tolerance is distinguished from design tolerance.

Process under control. A process in which the various factors affecting variability are maintained within defined control limits.

Producer's risk. The probability of rejecting a good batch from a producer's production lot.

QA. Quality Assurance.

QC. Quality Control.

QFD. Quality Function Deployment.

Qualification. The status given to an entity or person when the fulfillment of specified requirements has been demonstrated; the process of obtaining that status.

Qualification (of supplier). The process by which products are obtained from a supplier and tested for the purpose of documenting them as qualified or certified products. It is alternately referred to as *certification*.

Quality. The combined features and characteristics of a product that influence its ability to satisfy specified needs.

Quality assurance. The process that sets the standards for product quality. All those planned and systematic actions necessary to provide adequate confidence that a product or service will satisfy given quality requirements.

Quality audit. A systematic and independent examination to determine whether quality activities and results comply with planned arrangements and whether these arrangements are effectively implemented and are suitable to achieve objectives.

Quality characteristic. An aspect of an item that can be measured or observed with respect to how it contributes to the acceptability and/or functioning of the item.

Quality control. The operational techniques and activities that are used to satisfy quality requirements.

Quality manual. A document stating the quality policy, quality system, and quality practices of an organization.

Quality plan. A document setting out the specific quality practices, resources, and activities relevant to a particular product, process, service, contract, or project.

Quality policy. The overall intentions and direction of an organization regarding quality, as formally expressed by top management.

Quality potential. The probability that the values of the characteristics of a product will lie within specified limits.

Quality surveillance. The continuing monitoring and verification of the status of procedures, methods, conditions, products, processes, and services and the analysis of records in relation to stated references to ensure that requirements for quality are being met.

Quality system. The organizational structure, responsibilities, procedures, processes, and resources for implementing quality management.

Quality system audit. A documented activity performed to verify, by examination and evaluation of objective evidence, that applicable elements of the quality system are suitable and have been developed, documented, and effectively implemented in accordance with specified requirements.

Quality system review. A formal evaluation by management of the status and adequacy of the quality system in relation to quality policy and/or new objectives resulting from changing circumstances.

R chart (range chart). A control chart in which the subgroup range R is used for evaluating the stability of the variability of a process.

Random sampling. The process of selecting sample units in such a manner that all units under consideration have an equal chance of being selected as the sample.

Range. The difference between the smallest and largest values in a set of observations.

Range chart. The part of a quality control chart on which are plotted the values of the ranges of samples to provide a measure of the variability of the product and/or process.

Rational subgroup. One of the small groups within which it is believed that assignable causes are constant and into which observations can be subdivided in carrying out statistical analysis.

Rejects. The items of product that are not accepted because they fail to meet specific quality criteria.

Relative frequency. The ratio of the number of times a particular value (or a value falling within a given class) is observed to the total number of observations.

Repeatability. An indication of the closeness of the agreement between the results of successive measurements of the same value of the same physical quantity carried out under identical conditions.

Replication. The performance of an experiment or part of an experiment more than once. Each performance, including the first one, is called a replicate.

Response. The result obtained when an experiment is run under a specific set of conditions.

Rework. Any process whereby defective material is altered in an effort to make it acceptable.

Root cause. A fundamental deficiency that results in a nonconformance and must be corrected to prevent recurrence of the same or similar nonconformance.

Run. The process of producing a quantity of a product in a continuing sequence of operations within one production cycle.

Run chart. A graph of a characteristic versus sampling sequence used to detect trends.

s chart (sample standard deviation chart). A control in which the subgroup standard deviation is used for evaluating the stability of the variability of a process.

Sample. One or more units of a product drawn from a specific lot for the purpose of inspection.

Sample size. The number of units contained in a sample drawn from a production batch.

Sampling frequency. The ratio of the number of units of a product randomly selected for inspection at an inspection station to the number of units of the product going through the inspection station.

Sampling interval. The fixed interval of time or units of output between samples.

Sampling plan. A plan according to which one or more samples are drawn from a population.

Sampling procedure. The specific steps by which a sampling plan is carried out.

Scatter diagram. A plot of one variable against another that displays their relationship.

Scrap. A nonconforming unit of a product that is not usable and cannot be economically reworked.

Significance test. A statistical procedure to determine whether some quantity that is subject to a random variation differs from a hypothesized value by an amount greater than that attributable to random variation alone.

Skewed distribution. An asymmetric curve of a distribution having a longer tail to the right (skewed to the right) or to the left (skewed to the left).

SPC. Statistical Process Control.

Specification. A specification of the requirements to be met by a product for the product to be acceptable. The document that prescribes the requirements with which the product or service must conform.

Standard. The documented result of a particular standardization effort approved by a recognized authority.

Standard deviation. A measure of the dispersion of a set of values around its average value. When standard deviation is denoted by "s," it represents the sample standard deviation. When denoted by "σ," it represents the population or universe standard deviation.

Standard error. The standard deviation of a sampling distribution.

Standardization. The reduction of the number of characteristics or features of a system or the reduction of the number of ways these may vary or interact.

Statistic. A quantity calculated from a sample of observations used to establish an estimate of some population parameter.

Statistical control. The condition describing a process from which all assignable causes of variation have been eliminated and only chance causes remain. This is identified on a control chart by the absence of points beyond the control limits and by the absence of nonrandom patterns or trends within the control limits.

Statistical process control. The use of statistical tools such as histograms, control charts, and other variation analysis techniques to analyze a process or its output in order to take appropriate action to achieve and maintain a state of statistical control.

Statistical tolerance limits. A set of limits calculated from the results of sample observations and between which a stated fraction of the population will lie with a given probability.

Stratification. The physical or conceptual division of a population into separate parts called strata.

Subgroup. A set of elements having one or more characteristics in common.

Survey. An examination for some specific purpose; to inspect or consider carefully; to review in detail. Survey and audit are sometimes used interchangeably. Audit implies the existence of some agreed-upon criteria against which the plans and execution can be checked. Survey implies the inclusion of matters not covered by agreed-upon criteria.

System. A group of elements having dependent and independent effects that act together to achieve a specific function.

TBP. Team-Based Performance.

TEI. Total Employee Involvement.

Test. An examination of one or more characteristics of a product.

Test procedure. A measurement instruction describing the method by which one or more quality characteristics are to be assessed.

Testing. A means of determining an item's capability to meet specified requirements by subjecting the item to a set of physical, chemical, environmental, or operating actions and conditions.

Tolerance. The total allowable variation around a level or state (upper limit minus lower limit).

Tolerance limits. Limits that define the conformance boundaries for an individual unit of a product.

Total process variability. The inherent process variability plus variations from factors that have been allowed to change, such as operator errors, equipment adjustments, and so on.

TPM. Total Preventive Maintenance.

TQC. Total Quality Control.

TQM. Total Quality Management.

Traceability. The ability to trace the history, application, or location of an item or activity by means of recorded identification.

Treatment. A given combination of the levels of all factors to be included in an experimental study.

Triple C. A management concept that emphasizes the integration of communication, cooperation, and coordination functions for the purpose of improving process performance.

***u* chart (count per unit chart).** A control chart for evaluating the stability of a process in terms of the average count of a given classification event occurring within a sample.

Upper control limit (UCL). An upper limit of a range of values in a control chart used for determining when corrective actions may be needed.

Variable. A quantity that may take any one of a specified range of values.

Variance (population). A measure of dispersion of a population.

Variance (sample). An estimate of the measure of dispersion of a finite population based on a sample drawn from the population.

Verification. The act of reviewing, inspecting, testing, checking, auditing, or otherwise establishing and documenting whether items, processes, services, or documents conform to specified requirements.

X-bar chart (average chart). A control chart in which the subgroup average is used for evaluating the stability of the process level.

ZIPS. Zero Inventory Production System.

ZUD. Zero Unplanned Downtime.

APPENDIX D
UNITS AND MEASURES CONVERSION FACTORS

Number Prefixes

Prefix	SI Symbol	Multiplication Factors	Example
tera	T	$1\,000\,000\,000\,000 = 10^{12}$	tera fortune
giga	G	$1\,000\,000\,000 = 10^{9}$	giga byte
mega	M	$1\,000\,000 = 10^{6}$	mega bucks
kilo	k	$1\,000 = 10^{3}$	kilo byte
hecto	h	$100 = 10^{2}$	hectogram
deca	da	$10 = 10^{1}$	decade
deci	d	$0.1 = 10^{-1}$	decimal
centi	c	$0.01 = 10^{-2}$	centimeter
milli	m	$0.001 = 10^{-3}$	millimicron
micro	μ	$0.000\,001 = 10^{-6}$	microcomputer
nano	n	$0.000\,000\,001 = 10^{-9}$	nanosecond
pico	p	$0.000\,000\,000\,001 = 10^{-12}$	picosecond
femto	f	$0.000\,000\,000\,000\,001 = 10^{-15}$	femto chance
atto	a	$0.000\,000\,000\,000\,000\,001 = 10^{-18}$	atto likelihood

Area

Multiply	by	to obtain
acres	43,560	sq feet
	4,047	sq meters
	4,840	sq yards
	0.405	hectare
sq cm	0.155	sq inches
sq feet	144	sq inches
	0.09290	sq meters
	0.1111	sq yards
sq inches	645.16	sq millimeters
sq kilometers	0.3861	sq miles
sq meters	10.764	sq feet
	1.196	sq yards
hectare	10000	sq meters
sq miles	640	acres
	2.590	sq kilometers

Volume

Multiply	by	to obtain
acre-foot	1233.5	cubic meters
cubic cm	0.06102	cubic inches
cubic feet	1728	cubic inches
	7.480	gallons (US)
	0.02832	cubic meters
	0.03704	cubic yards
liter	1.057	liquid quarts
	0.908	dry quart
	61.024	cubic inches
gallons (US)	231	cubic inches
	3.7854	liters
	4	quarts
	0.833	British gallons
	128	U.S. fluid ounces
barrel	40	gallons
quarts (US)	0.9463	liters

Mass

Multiply	by	to obtain
carat	0.200	cubic grams
grams	0.03527	ounces
kilograms	2.2046	pounds
ounces	28.350	grams
pound	16	ounces
	453.6	grams
stone (UK)	6.35	kilograms
	14	pounds
ton (net)	907.2	kilograms
	2000	pounds
	0.893	gross ton
	0.907	metric ton
ton (gross)	2240	pounds
	1.12	net tons
	1.016	metric tons
tonne (metric)	2,204.623	pounds
	0.984	gross ton
	1000	kilograms

Temperature

Conversion formulas

Celsius to Kelvin	K = C + 273.15
Celsius to Fahrenheit	F = (9/5)C + 32
Fahrenheit to Celsius	C = (5/9)(F − 32)
Fahrenheit to Kelvin	K = (5/9)(F + 459.67)
Fahrenheit to Rankin	R = F + 459.67
Rankin to Kelvin	K = (5/9)R

Energy, Heat, Power

Multiply	by	to obtain
BTU	1055.9	joules
	0.2520	kg-calories
watt-hour	3600	joules
	3.409	BTU
HP (electric)	746	watts
BTU/second	1055.9	watts
watt-second	1.00	joules

Velocity

Multiply	by	to obtain
feet/minute	5.080	mm/second
feet/second	0.3048	meters/second
inches/second	0.0254	meters/second
km/hour	0.6214	miles/hour
meters/second	3.2808	feet/second
	2.237	miles/hour
miles/hour	88.0	feet/minute
	0.44704	meters/second
	1.6093	km/hour
	0.8684	knots
knot	1.151	miles/hour

Pressure

Multiply	by	to obtain
atmospheres	1.01325	bars
	33.90	feet of water
	29.92	inches of mercury
	760.0	mm of mercury
bar	75.01	cm of mercury
	14.50	pounds/sq inch
dyne/sq cm	0.1	N/sq meter
dyne	0.00001	Newton
newtons/sq cm	1.450	pounds/sq inch
pounds/sq inch	0.06805	atmospheres
	2.036	inches of mercury
	27.708	inches of water
	68.948	millibars
	51.72	mm of mercury

Length

Multiply	by	to obtain
angstrom	10^{-10}	meters
feet	0.30480	meters
	12	inches
inches	25.40	millimeters
	0.02540	meters
	0.08333	feet
kilometers	3280.8	feet
	0.6214	miles
	1094	yards
meters	39.370	inches
	3.2808	feet
	1.094	yards
miles	5280	feet
	1.6093	kilometers
	0.8694	nautical miles
millimeters	0.03937	inches
nautical miles	6076	feet
	1.852	kilometers
yards	0.9144	meters
	3	feet
	36	inches

Constants

speed of light	$2.997,925 \times 10^{10}$ cm/sec
	983.6×10^{6} ft/sec
	186,284 miles/sec
velocity of sound	340.3 meters/sec
	1116 ft/sec
gravity (acceleration)	9.80665 m/sec square
	32.174 ft/sec square
	386.089 inches/sec square

SELECTED BIBLIOGRAPHY

Akao, Y., and Asaka, T. (eds.). *Quality Function Deployment,* Productivity Press, Inc., Cambridge, MA, 1990.

Badiru, Adedeji B. "Communication, Cooperation, Coordination: The Triple C of Project Management," in *Proceedings of 1987 IIE Spring Conference,* Washington, DC, May 1987, pp. 401–404.

Badiru, Adedeji B. *Project Management in Manufacturing and High Technology Operations,* John Wiley & Sons, New York, 1988.

Badiru, Adedeji B. "A Systems Approach to Total Quality Management," *Industrial Engineering,* Vol. 22, No. 3, March 1990, pp 33–36.

Badiru, Adedeji B. "Systems Integration for Total Quality Management," *Engineering Management Journal,* Vol. 2, No. 3, Sept. 1990, pp 23–28.

Badiru, Adedeji B. *Project Management Tools for Engineering and Management Professionals,* Industrial Engineering & Management Press, Norcross, GA, 1991.

Badiru, Adedeji B. "Total Quality Management: A Project Management Approach," *Proceedings of Project Management Institute Annual Symposium,* Dallas, September 1991, pp 62–67.

Badiru, Adedeji B. *Expert Systems Applications in Engineering and Manufacturing,* Prentice-Hall, Englewood Cliffs, NJ, 1992.

Badiru, Adedeji B. *Managing Industrial Development Projects: A Project Management Approach,* Van Nostrand Reinhold, New York, 1993.

Badiru, Adedeji B., and Ayeni, B. J. *Practioner's Guide to Quality and Process Improvement,* Chapman & Hall, London, 1993.

Badiru, Adedeji B., and Chen, Jacob Jen-Gwo. "IEs Help Transform Industrial Productivity and Quality in Taiwan," *Industrial Engineering,* Vol. 24, No. 6, June 1992, pp 53–55.

Badiru, Adedeji B., and Pulat, P. S. *Comprehensive Project Management: Integrating Optimization Models, Management Practices, and Computers,* in-press, Prentice-Hall, Englewood Cliffs, NJ, 1995.

Crosby, Phillip B. *Quality if Free: The Art of Making Quality Certain,* McGraw-Hill, New York, 1979.

Crosby, Phillip B. *Quality Without Tears: The Art of Hassle-free Management,* McGraw-Hill, New York, 1984.

Deming, W. Edwards. *Quality, Productivity and Competitive Position,* MIT, Center for Advanced Engineering Study, Cambridge, MA, 1982.

Dingus, Victor and Golomski, W. (eds.). *A Quality Revolution in Manufacturing,* Industrial Engineering & Management Press, Norcross, GA, 1991.

Ebrahimpour, M. and Withers, Barbara E. "Employee Involvement in Quality Improvement: A Comparison of American and Japanese Manufacturing Firms Operating in the U.S.," *IEEE Transactions on Engineering Management,* Vol. 39, No. 2, May 1992, pp. 142–148.

Enrick, Norbert L., *Quality, Reliability, and Process Improvement,* 8ed, Industrial Press, Inc., New York, 1985.

Feigenbaum, A. V., *Total Quality Control,* McGraw-Hill, New York, 1983.

Gitlow, Howard; Gitlow, Shelly, Oppenheim, Alan; and Oppenheim, Rosa. *Tools and Methods for the Improvement of Quality,* Irwin, Homewood, IL, 1989.

Hirano, Hiroyuki. *JIT Implementation Manual: The Complete Guide to Just-in-Time Manufacturing,* Productivity Press, Inc., Cambridge, MA, 1991.

Hordes, Mark D. "10 Burning Questions Concerning Quality Improvement," *Industrial Engineering,* Vol. 24, No. 9, 1992, pp 56–57.

Hradesky, John L. *Productivity and Quality Improvements,* McGraw-Hill, New York, 1988.

Huyler, George and Crosby, K. "The Best Investment a Project Manager Can Make: Improve Meetings," *PM Network,* Vol. 7, No. 6, June 1993, pp 33–35.

Imai, Masaaki. *Kaizen: The Key to Japan's Competitive Success,* Random House, New York, 1986.

Ishikawa, Kaoru. *Guide to Quality Control,* 2nd rev. ed., Quality Resources, White Plains, NY, 1986.

Ishikawa, Kaoru. *Introduction to Quality Control,* Quality Resources, White Plains, NY, 1990.

Johnson, Perry. *ISO 9000: Meeting the New International Standards,* McGraw-Hill, New York, 1993.

Juran, J. M. *Juran on Planning for Quality,* The Free Press, New York, 1988.

Juran, J. M. *Juran on Leadership for Quality,* The Free Press, New York, 1989.

Juran, J. M., and Gryna, F. M. *Juran's Quality Handbook,* 4th ed., McGraw-Hill, New York, 1980.

Juran, J. M., and Gryna, Frank M. Jr. *Quality Planning and Analysis,* 2nd ed. McGraw-Hill, New York, 1980.

Kantner, Rob. *The ISO 9000 Answer Book,* Industrial Engineering & Management Press, Norcross, GA, 1994.

Kelly, Michael R. *Everyone's Problem Solving Handbook: Step-by-Step Solutions for Quality Improvement,* Quality Resources, White Plains, NY, 1991.

Koulamas, C. "Quality Improvement Through Product Redesign and the Learning Curve," *Omega,* Vol. 20, No. 2, 1992, pp. 161–168.

Kume, Hitoshi. *Statistical Methods for Quality Improvement,* AOTS, Tokyo, 1989.

Lamprecht, J. L. *Implementing the ISO 9000 Series,* Marcel Dekker, New York, 1993.

Lehr, Lewis. "Quality is a Measure of Company Success," *IE Financial Services News,* Institute of Industrial Engineers, Norcross, Georgia, Vol. XXIII, No. 2, Winter 1989, p.1.

Lester, Ronald H.; Enrick, N. L.; and Mottley, H. E. *Quality Control for Profit: Gaining the Competitive Edge,* 3rd ed., Marcel Dekker, Inc., New York, 1992.

Lillrank, Paul, and Kano, Noriaki. *Continuous Improvement: Quality Control Circles in Japanese Industry,* Center for Japanese Studies, The University of Michigan, Ann Arbor, MI, 1989.

Lochner, Robert H., and Mater, Joseph E. *Designing for Quality: An Introduction to the Best of Taguchi and Western Methods of Statistical Experimental Design,* Quality Resources, White Plains, NY, 1990.

Maslow, Abraham H. *Motivation and Personality,* Harper & Brothers, New York, 1954.

Mizuno, Shigeru. *Company-wide Total Quality Control,* Asian Productivity Organization, Tokyo, 1988.

Mizuno, Shigeru, editor, *Management for Quality Improvement,* Quality Resources, White Plains, NY, 1988.

Omachonu, Vincent K. *Total Quality and Productivity Management in Health Care Organizations,* Industrial Engineering & Management Press, Norcross, GA, 1991.

Ott, E. R. *Process Quality Control,* McGraw-Hill, New York, 1975.

Ott, E. R., and Schilling, E. G. *Process Quality Control: Troubleshooting and Interpretation of Data,* 2nd ed., McGraw-Hill, New York, 1990.

Persico, John, Jr., (ed.). *The TQM Transformation: A Model for Organizational Change,* Quality Resources, White Plains, NY, 1992.

Rabbitt, John T., and Bergh, P.A. *The ISO 9000 Book: A Global Competitor's Guide to Compliance & Certification,* Quality Resources, White Plains, NY, 1993.

Ross, Phillip J. *Taguchi Techniques for Quality Engineering,* McGraw-Hill, New York, 1988.

Russell, J. P. *Quality Management Benchmark Assessment,* Quality Resources, White Plains, New York, 1991.

Ryan, Thomas P. *Statistical Methods for Quality Improvement,* John Wiley & Sons, New York, 1989.

Scherkenback, W. W. *The Deming Route to Quality and Productivity: Roadmaps and Roadblocks,* Mercury Press/Fairchild Publications, New York, 1990.

Schonberger, Richard J. *Just In Time: A Comparison of Japanese and American Manufacturing Techniques,* Industrial Engineering & Management Press, Norcross, GA, 1984.

Somasundaram, S., and Badiru, Adedeji B. "Project Management for Successful Implementation of Continuous Quality Improvement," *International Journal of Project Management,* Vol. 10, No. 2, May 1992, pp. 89–101.

Son, Y. K., and Park, C. S. "Economic Measure of Productivity, Quality, and Flexibility in Advanced Manufacturing Systems," *Journal of Manufacturing Systems,* Vol. 6, 1987, pp. 193–206.

Steeples, M. *The Corporate Guide to the Malcolm Baldrige National Quality Award,* Business-One Irwin, Homewood, IL, 1992.

Stewart, D. L. "Management Meetings Could be New Career Field," *The Norman Transcript,* April 25, 1993, p. 10A.

Taguchi, Genichi. *Introduction to Quality Engineering: Designing Quality into Products and Processes,* Asian Productivity Organization, Tokyo, 1990.

Wadsworth, H. M.; Stephens, K. S.; and Godfrey, A. B. *Modern Methods for Quality Control and Improvement,* John Wiley & Sons, New York, 1986.

Walton, Mary. *The Deming Management Method,* Perigee Books/Putnam Publishing Group, New York, 1986.

Willborn, Walter. *Quality Management System: A Planning and Auditing Guide,* Industrial Press, Inc., New York, 1989.

Wilson, M. "New Paradigms for Project Plans," *PM Network,* Vol. 7, No. 6, June 1993, pp. 39–40.

Winchell, B. *Continuous Quality Improvement: A Manufacturing Professional's Guide,* Society of Manufacturing Engineers, Dearborn, MI, 1991.

Winchell, B., (ed.), *TQM: Getting Started and Achieving Results with Total Quality Management,* Society of Manufacturing Engineers, Dearborn, MI, 1992.

INDEX

A
Agreements, trade, 4
American National Standards Institute (ANSI), 18, 22
American Society for Quality Control (ASQC), 22
ANSI/ASQC Q90 series, 18, 22
American Registrar Accreditation Board, 31
Audit:
 external, 151
 internal, 29–31, 149–168
  checklist for, 29–31
  purpose of, 149–150
  types of, 150–151
   process, 151
   product, 151
   system, 150–151
 paper, 38, 40, 159–168
 sampling strategy, 152–157
 team preparation, 152
 third-party, 31–33
Auditor:
 internal, 12–13
 lead, 12

B
Badiru, A. B., 6, 7
Baldrige award, 5
Benchmarking, 8–9
 definition of, 8
Blueprint, for project, 109
Brainstorming, 101
BS 5750, 18
Budget planning, 128–129
 bottom-up, 129, 133
 top-down, 128

C
Cause-and-effect analysis, 68
Certificate of Registration, 31
Certification, vendor, 51–52
Change:
 preparation for, 6
 technological, 43–44
Committee:
 ad hoc, 94
 standing, 94
 steering, 61
Communication, 135–139
 barriers to, 137
 matrix, 139

Competition, global, 2, 6
Complexity, product, 55–57
Compliance schedule, 41
Conflict resolution, 145–147
Continuous flow manufacturing, 48
Continuous measurable improvement (CMI), 63
Continuous process improvement (CPI), 20, 60–63
  advantages of, 63
Contract:
  requirements, 121
  review of, 72, 160–161
Control, project, 118
Cooperation, 140–142
CPM, 89, 112
Crosby, Phillip B., 100
Customer involvement, 48–51

D
Data:
  measurement scales, 153–154
  types of, 153
Decision making, group, 100–106
  methods for, 101–106
Degradation-innovation cycle, 61–62
Delphi method, 101–103
Deming, W. Edwards, 48, 58
Design:
  changes, 161
  control of, 72–73, 160
  input, 160
  output, 160–161
  planning, 160
Documentation, 33–41
  changes and modifications, 162–163
  control of, 73, 162
  coordinator, 13, 38
  hierarchy, 34
  levels, 35–36
  paper trail for, 37–38
  process steps, 34–35
  survey, 36, 39, 41

E
Employee involvement, 54–55
EN 29000, 18
Errors:
  sampling and nonsampling, 155
  Type I, 13–14
  Type II, 13–14

F
Feasibility study, 126–128

G
Gantt charts, 89, 112
GATT, 4–5
Goals, project, 117

H
Huyler, George, 100

I
Inspection, 25
  equipment for, 77–78
  and testing, 76–77, 78, 164–166
  records, 165–166
International Organization for Standardization (IOS), 1
International Standards Organization (ISO), 1
  objective of, 18
  policy and procedures, 41
Interval scale, 153
Interviews, 105, 157–159
ISO 9000:
  acceptance of, 25–27
  benefits of, 18–19, 26–27
  coordinator, role of, 12
  definition of, 17–18
  do's and don'ts of, 43
  importance in world trade, 4
  information flow for, 131–134
  information sources, 169–171
  leadership model, 10
  meetings, 97–100
  mission logistics, 14–15

INDEX **211**

organizational structure, 11
plan, components of, 120–122
planning guidelines, 115–147
project decision process, 94–97
   data and information
     requirements, 95
   decision model, 96
   project outline, 110–113
purpose of, 21
registration process, 28–29
related standards to, 18
subdivisions of, 20–21
systems integration for, 107–109
   flowchart, 109
tips for successful process, 42–43
total improvement through, 68
ISO 9001, 20, 28–29, 31, 35, 36, 37, 152
   quality system requirements, 67–81
ISO 9002, 20, 31
ISO 9003, 20, 31
ISO 9004, 21, 36

J
Job descriptions, 40–41
Job ownership, 55
Job transfer, 57
Just in time (JIT), 47–48
   benefits of, 48

K
Kanban, 48
KISS (keep it simple, sweetie), 42

M
MAN (material as needed), 47
Management:
   by exception (MBE), 58–59
   leadership, 9–13
   by objective (MBO), 58–59
   by project (MBP), 83–86
   responsibilities of, 10–11, 25, 70–71
   review, 151–152, 160
   role of representative, 11–12, 71

Market research, 64
Maslow, Abraham, 123
Massachusetts Quality Award, 5
Minnesota Quality Award, 5
MIPS (minimum inventory production system), 47
Mission statement, 15, 87
Motivation of employees, 122–125
   factors in, 124–125
   needs hierarchy, 123
Multivoting, 105

N
NAFTA, 4
Nominal group technique, 103–104
Nominal scale, 153

O
Ohno system, 48
Oklahoma State Quality Award, 5
Ordinal scale, 153
Overhead, allocation of, 56

P
Pareto distribution, 153
Partnership, producer-consumer, 49–50
PDCA cycle, 48
Performance, 117–118
   measures of, 121
   trade-offs with time and resources, 118, 119
Personnel:
   management, 106–107
   responsibilities, 159
PERT, 89, 112
Plan, selling of, 91–92
Planning:
   levels, 6–7
     macro-level, 7
     micro-level, 7
     supra-level, 7
   as part of project outline, 110
   of product, 56
   strategic, 7, 116–117

INDEX

Process control, 75, 163–164
Product:
  characteristics, 56
  complexity, 55–56
  configuration, 56
  delivery of, 80, 167
  handling of, 79, 167
  identification and traceability, 75, 163
  nonconforming, 38, 78–79
    control of, 166
    corrective action for, 79, 166–167
  packaging of, 79, 167
  purchaser-supplied, 74
  storage of, 79, 167
  tracking, 57
Production, stockless, 48
Project breakdown structure, *see* Work breakdown structure
Project management, 83–113
  definition of, 83
  elements of, 83–86
  functional areas, 84–86
  steps of, 86–89
Project manager, selection of, 90–91
Purchasing, 74, 162

Q

Quality:
  awards for, 5–6
  definitions of, 2, 49
  improvement:
    components of, 48
    flowchart, 49
  language of, 54
  manual, 71
  of manufactured goods, 55–57
  policy, 70, 159
  problems, prevention of, 60
  proclamation, 17
  records, 80, 167
  of service, 57
Quality assurance, 21
  fishbone diagram for, 69
  standards, table of, 23–24
Quality circles, 54
Quality function deployment (QFD), 64
Quality system, 22, 160
  characteristics of, 24
  documentation, 25, 71
  elements of, 22
  integration, 50
  objective of, 24
  requirements of, 25
Questionnaires, 105

R

Ratio scale, 154
Registered Firm Symbol, 31
Reporting, 112
Resources:
  allocation of, 88, 112, 133
  requirements, 121, 133
Responsibility matrix, 28, 36, 41, 139, 142–143, 144
Risk:
  consumer's, 13
  management of, 13–14
  producer's, 13

S

Sampling:
  bias, 155
  characteristics, 154
  cluster, 156
  errors, 155
  multistage, 157
  stratified, 156
  types of, 155–157
Scheduling, 88–89, 112
  techniques, 89, 112
Servicing:
  procedures for, 168
  specified in contract, 81
Site implementation plan, 41
Staffing, 92–94
Standards:
  consensus, 3
  contractual, 3

international, 3–4
regulatory, 3
Standards Organization of Nigeria, 5
Statistical techniques, 81
Stewart, D. L., 98
Surveys, 105
  customer, 64

T
Teamwork, 7–8, 104, 143, 145
Testing, *see* Inspection
Theory X and Theory Y, 123
Timeline, 42
Total employee involvement (TEI), 54–55
Total quality management (TQM), 20, 45–66
  benefits of, 47
  case study, 65–66
  role of management in, 59
  systems integration, 45–46
Toyota system, 48
Tracking, 112

Training:
  personnel, 80–81, 168
  user, 51
Triple C model, 28, 111, 134–147
  communications matrix, 139

U
Underwriters Laboratory (UL), 31

V
Vendor:
  certification, 51–52
  rating matrix, 53
  rating system, 52–54
Verification activities, 70, 160

W
Win-win approach, 7
Work breakdown structure (WBS), 130–131

Z
ZIPS (zero inventory production system), 47